Coral Fish

Linda Pitkin

SMITHSONIAN INSTITUTION PRESS, WASHINGTON, D.C.
IN ASSOCIATION WITH THE NATURAL HISTORY MUSEUM, LONDON

For my husband, Brian

Published in the United States of America
by the Smithsonian Institution Press in association with
The Natural History Museum
London
Cromwell Road
London SW7 5BD
United Kingdom

Library of Congress Cataloging-in-Publication Data
Pitkin, Linda M.
    Coral fish / Linda Pitkin.
        p.   cm.
    Includes bibliographical references (p.).
    ISBN 1-56098-818-5 (alk. paper)
    1. Coral reef fishes. 2. Coral reef ecology. I. Title.

QL620.45.P58   2001
597.177'89—dc21   00-046320

Manufactured in Singapore, not at government expense
08 07 06 05 04 03 02 01 5 4 3 2 1

DISTRIBUTION

United States, Canada, Central and
South America, and the Caribbean
Smithsonian Institution Press
470 L'Enfant Plaza
Washington, D.C. 20560-0950
USA

Australia and New Zealand
CSIRO Publishing
PO Box 1139
Collingwood, Victoria 3066
Australia

UK and rest of the world
Plymbridge Distributors Ltd.
Plymbridge House, Estover Road
Plymouth, Devon PL6 7PY
UK

Edited by Jacqui Morris
Designed by Mercer Design
Reproduction and printing by Craft Print, Singapore

On the front cover: Scales of queen angelfish,
Holacanthus ciliaris; inset, Basslets of the
subfamily Anthiinae.
On the back cover and title page: Golden sergeant,
Amblyglyphidodon aureus.

Contents

Preface

Coral reefs occupy less than 1% of the Earth's crust, but they support an immense diversity of fishes throughout the tropics. These include over 4000 species, ranging in size from a giant grouper, 3 m (9.8 ft) long, to a minute goby, barely 1 cm (0.4 in) in length. The obvious appeal of coral fishes lies in the colour and beauty of many of the more familiar ones, such as butterflyfishes and angelfishes, while tiny damselfishes, and others that may be less conspicuous as individuals, shimmer in lively swarms around the corals. The fascination of coral fishes, however, goes far beyond first impressions; get to know them better and they reveal amazing lifestyles and behaviour – sex change, harem maintenance and egg-brooding by the male, to name a few of the strategies that some of them exhibit.

This book explores some of the varied specializations that enable coral fishes to live in every part of the reef environment – the means by which they derive food and shelter from the reefs, and their interactions, often fiercely territorial, in a densely populated living space. Many coral fishes are secretive, hiding in holes or burrows, or lying camouflaged on the sea bed; others hover openly in the water column above. Their impact on coral reefs is immense, whether they graze the fine seaweeds growing on the reef, crunch the coral, or prey on various reef-dwelling animals. Sharks, schooling snappers and jacks are not usually thought of as coral fishes but some of their members live on or around coral reefs and are significant predators of reef fishes, and these are therefore included too.

Author

Linda Pitkin is a biologist with two main fields of interest. By profession she is an entomologist at The Natural History Museum in London, specializing in research on moths. Aside from this, her greatest passion is observing and photographing marine life. Her underwater photographs have won many national and international awards; they have been widely published in books and magazines, and have also featured in an IMAX multi-screen presentation. She is the author of *The Living Sea – a Photographic Exploration of Life in the Sea* (1995), and *Under Northern Seas* (1997).

The coral reef environment

Coral reefs are limestone structures formed by the skeletons of tiny sea-anemone-like animals – corals belonging to the order Scleractinia. Reef-building corals live in colonies, which vary immensely in growth form but have one feature in common: they all contain zooxanthellae (single-celled algae) in their tissues. These minute algae play a vital role in reef construction. Like most green plants, they produce carbohydrates by photosynthesis, and it is this supply of nutrients that not only keeps their coral partners alive and well, but enables them to lay down rapid deposits for skeletal growth.

LEFT **Coral polyps – the individual animals of the colony – of a Caribbean star coral *Montastrea* sp.**

LEFT **A coral reef community in the Indo-Pacific.**

Coral-reef communities

A coral reef is a community of many species: corals, other animals and plants. They all have their preferred sites, some colonizing the shallow zones and the outer reef, exposed to surge and currents, and others living in deeper or more sheltered regions. Even when the same species of coral lives in more than one zone, it may not look the same at different sites. Depth, light, water movement, temperature and water clarity are among the conditions that vary in different parts of a reef and affect the way in which a coral colony grows. Many corals are naturally long-lived, but they are intolerant of changes in their surroundings. Life on a reef is abundant in the shallows, but reef-building corals cannot live deeper than about 50 m (164 ft) because their zooxanthellae need sunlight. Corals require a water temperature of 18°C or more for reef formation and coral reefs are thus essentially tropical. They are widespread in two main regions, the Indo-Pacific region from the Red Sea and South Africa to Polynesia, and the Western Atlantic region, the chief component of which is the Caribbean.

The coral reef ecosystem is highly complex, containing an amazing diversity of invertebrates (animals without backbones), fishes, and some other vertebrate animals, such as turtles. Although reefs cycle some nutrients to and from other environments, they are to a large extent self-contained systems, and are densely populated with inhabitants. Individuals of different species and within species interact with each other in various ways, as predators, prey, competitors, mates or co-operative partners. Studies and observations have revealed fascinating aspects of their lifestyles but much remains unknown, and many secretive small fishes and other animals are still being discovered.

BELOW **A snapper among gorgonian coral plumes on a Caribbean reef.**

Coral fish diversity

Coral fish communities are extremely diverse and dense on many tropical reefs, more so than in any other aquatic habitat. Over 4000 fish species – nearly one-third of all known marine fish species – have been recorded on reefs. Some families of fishes are adapted to live primarily on coral reefs – butterflyfishes and damselfishes, for example – while others, such as wrasses, have many members living in other habitats. Coral fishes live not only among the reef-building corals but also with sea fans and soft corals, sponges and sea anemones. Some fishes rest on patches of sand or peep out of holes in the reef, others hover above the reef or swim actively, and visitors from the open ocean come in to prey on the residents.

How can a coral reef support so many fishes? This is possible by the diverse lives that the fishes lead, specializing in various foods, and occupying different zones and habitats on and around the reef. The pressures of

BELOW **Morays, such as the yellowmouth moray** *Gymnothorax nudivomer*, **occupy holes and passages in the reef.**

predation and competition are high and have given rise to immense variety in modes of life and behavioural ploys, and the physical adaptations needed to carry them out.

Coral fishes are not all active at the same time. Some feed during the daytime and retire to other sites to sleep at night, while others move out from daytime shelters to feed at night. Only a small minority are active on and off throughout day and night. When they are not feeding or breeding, most reef-dwellers hide from predators. Many fishes use the same refuges at different times of the day, making the most effective use of valuable space. Small fishes, such as blennies and gobies, do not tend to travel far, so their refuges and feeding sites are close together. Some large fishes, however, commute considerable distances, sometimes several kilometres between resting and feeding places. Many form schools for safety when they are travelling, so that each individual fish runs less risk of being singled out by predators.

BELOW **Bluestripe snappers** *Lutjanus kasmira* **shelter in groups by day and, like most other snappers, roam to feed at night.**

Major coral fish families

Coral fishes are classified in more than 100 different families. Some of these are found only on coral reefs and others have some members that live in other habitats as well. The vast majority, including all the major families listed below, belong in the bony fishes (i.e. fishes with a skeleton of bone) but a small minority are cartilaginous (fishes with a skeleton of cartilage). In the latter group are sharks and rays, a few species of which live on or around coral reefs.

The fish families listed below are either particularly characteristic of coral reefs or are important in terms of numbers of reef species.

Damselfishes & anemonefishes	Pomacentridae
Wrasses	Labridae
Parrotfishes	Scaridae
Surgeonfishes	Acanthuridae
Butterflyfishes	Chaetodontidae
Angelfishes	Pomacanthidae
Groupers & seabasses	Serranidae
Blennies	Blenniidae
Gobies	Gobiidae
Cardinalfishes	Apogonidae
Squirrelfishes & soldierfishes	Holocentridae
Grunts	Haemulidae

ABOVE **French angelfish** *Pomacanthus paru* **– a member of the family Pomacanthidae.**

RIGHT **Members of the grunt family are found on many coral reefs – grunts of the genus** *Haemulon* **are often abundant on reefs in the Caribbean.**

Food for coral fishes

Some coral fishes prey on other animals; others are omnivores or herbivores. To the casual observer, plants do not appear to be much in evidence on a coral reef, but they are there in abundance as fine growths of algae or seaweed on dead corals and rock surfaces, especially in the shallows. Parrotfishes, surgeonfishes and some damselfishes graze the algae, contributing to the diversity of a coral reef by clearing new surfaces that can be used by a wide range of colonizing animals.

When one thinks of predatory fishes, the top hunters around coral reefs – sharks and barracudas – usually spring to mind but the numerous smaller predators are equally or more significant. Nearly 1000 goby species can be found on coral reefs, catching mainly shrimps, worms and other tiny, bottom-living animals, while the water above the corals teems with damselfishes that eat the zooplankton suspended there. Almost all reef organisms – the coral itself and the many different invertebrate animals and fishes that the reef supports – are potential food for one

BELOW **Powder-blue surgeonfish** *Acanthurus leucosternon* **graze algae on the reef.**

Names

Many familiar fish are known by more than one common name, particularly in different countries. This applies to individual species and also to groups of fishes, and can cause confusion.

Scientific names, on the other hand, constitute a universal language that helps to overcome this problem. To give an example, the names lionfish, turkeyfish, red firefish, dragonfish and butterfly cod all refer to the same species of fish – *Pterois volitans* – and some of those common names are used also for various related species. (For further examples see p. 41.)

Classification of a fish – an example

Phylum	Chordata (animals with a notochord, including the vertebrates)
Class	Osteichthyes (bony fishes)
Order	Perciformes (perch-like fishes, including the majority of coral fishes)
Family	Scaridae (parrotfishes)
Genus	*Sparisoma* (a group of closely related parrotfish species)
Species	*viride* (stoplight parrotfish)

The fish is referred to by two names (genus and species) together – *Sparisoma viride*.

RIGHT **Glasseyes**
Heteropriacanthus
cruentatus **are members**
of the bigeye family,
which commonly feed on
plankton at night.

or other coral fish species. Small coral fishes are preyed on by larger fishes and also by certain other reef inhabitants, among them cone shells (venomous molluscs), mantis shrimps, and some starfishes.

Many fishes are able to feed during the day because their food – corals, seaweeds, and various animals attached to the reef – is available out in the open at all times. The major coral-feeding fishes are certain butterflyfishes, but parrotfishes and other herbivores may have a major impact on the underlying corals by cropping algae. Sponges and sea squirts are eaten, particularly by the larger angelfishes, but hydroids (animals with a fern-like appearance) are usually avoided because they have stinging cells. The omnivorous scrawled filefish *Aluterus scriptus*, a circumtropical species, however, includes hydroids in its diet. Fishes that feed on more active animals often wait until dusk or later, in order to pick off fishes as they scramble for night shelter, or to catch crabs, octopuses and other invertebrates emerging from their holes in the reef. Plankton – the speciality of soldierfishes, bigeyes and cardinals – is another major food source. It is available to daytime feeders but is more plentiful and safer to feed on after dark, when large numbers of planktonic animals rise up from deeper water.

Plankton-feeding: an alternative strategy

A healthy coral reef is rich in food for its inhabitants, not only on the reef itself but also in the water column above it. Plankton suspended in the water is an important source of nutrients and various coral fishes specialize in using it. They are not related to each other but many of them have similar features that go with the lifestyle. Daytime plankton-feeders are vulnerable, especially those that swim high up in the water column where they are obvious to predators and exposed to currents, and they need to be strong swimmers. Typically, their small bodies are streamlined and their tails are forked to enhance propulsion. The fishes tend to gather in large numbers for safety while feeding, and all dart for shelter if a cruising predator approaches. Catching tiny, fragile plankton requires other adaptations: a small mouth without the need for strongly developed teeth but with jaws capable of rapid and extensive protrusion to snatch the small morsels drifting by. In some cases the action of extending the jaws creates a suction force, drawing water and particles in.

Fishes adapted for plankton-feeding may look very different from typical members of their family. This is particularly noticeable when the norm for the family is a large build, as in groupers and snappers, but less apparent when the build is generally small, as in damselfishes. Many damselfishes, particularly the numerous pullers of the genus *Chromis*, are planktivores. Others include basslets (the subfamily Anthiinae of the grouper and seabass family), fusiliers (related to snappers and sometimes included in that family), and certain wrasses.

LEFT **Small plankton-feeding fishes of the genera *Pseudanthias* and *Dascyllus* hover above a table coral.**

Territorial behaviour

Living at such close quarters on a coral reef, many fishes guard their territories against competitors and other unwelcome intruders. Sites guarded vary according to the lifestyle of the fish; some have a permanent retreat, such as a goby's burrow or a moray's hole in the reef. Herbivorous damselfishes are fiercely protective of their feeding grounds and, like various other fishes, protect breeding and nesting sites. The first sign that a fish gives when its personal space is invaded is often to spread its fins, and sometimes also its gill covers. If this fails to intimidate, it will make excited movements and sometimes warning sounds, and, as a last resort, chase the intruder away.

Social colours

Why are coral fishes often so colourful when camouflage is safer? The answer lies primarily in their social behaviour. On a heavily populated reef, interactions between fishes are frequent and the right response is vital. Does the confrontation call for a territorial display or mating advances? A colour pattern that can be recognized rapidly is an advantage. Fishes that are active during the daytime and that live in the shallower zones have well-developed vision for a range of colours, sometimes including ultraviolet. In the clear water and abundant sunlight on the upper levels of a coral reef, bright colour patterns are visible for some distance, enabling a butterflyfish, for example, to distinguish its own kind immediately from the many others around. Colour is especially significant for many coral fishes at breeding

OPPOSITE **Like many other reef fishes, scalefin anthias *Pseudanthias squamipinnis* are small but conspicuously coloured.**

BELOW **The false clown anemonefish *Amphiprion ocellaris* lives with the magnificent sea anemone *Heteractis magnifica*, or with certain other anemone species.**

times; sexual differences are common in some groups, and heightened colours of males signal readiness to mate. Complex courtship rituals may even involve flashing eye-catching markings, usually on the fins, to woo a female. Fin flashing is common also in aggressive encounters, although deployed in a different manner. Rapid colour changes may occur in these situations and at other times, such as when a fish takes cover or prepares for the night.

Conspicuous markings may also be warning colours, labelling fishes as unsuitable to eat. Highly visible coral fishes need a line of defence against predators, and many of them have well-developed fin spines. Some

are armed with razor-like blades (surgeonfishes), or are poisonous (certain pufferfishes). Warning coloration, of toxic fishes in particular, is sometimes mimicked by species that lack defensive toxins. A notable example is the striking resemblance of the blacksaddle filefish *Paraluteres prionurus* to the poisonous toby or sharpnose pufferfish *Canthigaster valentini*. Many fishes are patterned with contrasting bands, which are not always as conspicuous as might be supposed; such markings can break up the body shape, making it less obvious to an onlooker.

Not all coral fishes are brightly coloured, but those that are make an immediate and lasting impression on the observer. Drab fishes, on the other hand, may escape notice or are quickly passed by, but they are even more numerous. They are everywhere on a reef, in particular on the sea bed, the home of many gobies, scorpionfishes and others of all shapes and sizes.

Co-operative relationships

Not all relationships between different species are competitive or threatening. Some are of benefit to one or both participants, as in the symbiosis (living together) of anemonefishes and sea anemones (p. 32–36), or prawn-gobies that share the burrows of blind prawns (p. 74). Long-term partnerships and temporary liaisons are many and varied on a coral reef. One that most coral fishes cannot do without is the relationship with cleaners: chiefly certain small Indo-Pacific wrasses and Caribbean gobies, and also some shrimps, which perform a useful service by eating

LEFT **The cleaner wrasse** *Labroides dimidiatus* **attends to a sabre squirrelfish** *Sargocentron spiniferum*.

external parasites, and dead or diseased skin. Cleaning stations are plentiful on a healthy reef and mostly serve a wide range of fish clients, although large oceanic visitors such as sharks and manta rays may have their own specific sites.

Reproduction and development

Coral fishes vary tremendously in reproductive behaviour. Their sex is not always fixed early: a change from female to breeding male is the norm in wrasses, parrotfishes and some groupers, while in anemonefishes the reverse change happens, from male to female. Some coral fishes bond in monogamous pairs, in others the male has a harem. Group mating and spawning occur in certain surgeonfishes and groupers.

Spawning often occurs late in the afternoon or at dusk, but also at other times for some fishes. The frequency of the event may be every other day, weekly, monthly, or less often, and in many cases it is synchronized with the phases of the moon. The majority of coral fishes rise up in pairs to spawn in midwater but some scatter their eggs on the sea bed. Others go in for a degree of parental care, usually by the male. Damselfishes, and some triggerfishes and gobies, guard nests on the sea bed, but only a few fishes go to the extreme of carrying the eggs around with them, in the mouth (jawfishes and cardinals) or in a pouch on the body (seahorses and pipefishes).

From hatchling to adult

Hatching from the eggs, the tiny larval fishes generally embark on a phase of life in open water away from the reef. During this time they form part of the plankton for between 10 to 100 days, the actual period depending on the species. They are transparent but extremely diverse in form, and often look amazingly different from the adults. Some larval fishes have various bizarre specializations such as long spines and filaments. Some are well developed as soon as they hatch, but many kinds of fishes start out still dependent on a sizable yolk sac. When that is used up they must feed on the minute planktonic animals around them.

Larval fishes may travel varying distances during this period. It is a means of dispersal to new territories, and probably safer than remaining close to the reef where predators abound, but even so the delicate hatchlings face many dangers in the open ocean. The survivors settle eventually on a reef (or a seagrass bed adjacent to it) as juvenile fishes.

BELOW **Larval fish are transparent and usually very unlike the adults in appearance.**

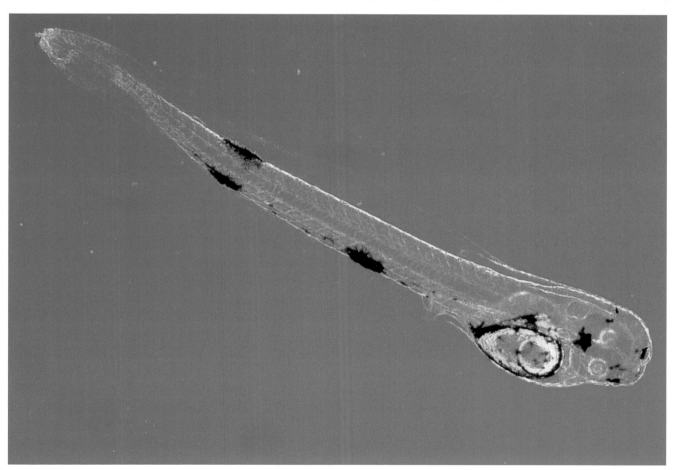

ABOVE **Juvenile batfish (spadefish)** *Platax teira* **are very deep-bodied, and their elongate fins gradually shorten as the fish become adult.**

They usually arrive at night under cover of darkness and remain extremely vulnerable to predation for the first few weeks until they have learnt the ropes. Some new arrivals will have already undergone transformation to develop juvenile features, but others will be still small and not yet opaque. When the juvenile stage is reached, an individual may look like a smaller version of the adult, but this is not always so. In some cases the colour pattern is different from the one they will acquire in maturity – a striking example is seen in angelfishes of the genus *Pomacanthus* (p. 28).

Fish-watching

There is much to appreciate in a well-maintained aquarium, but the best place to watch coral fishes is in their own environment, the coral reef. Diving and snorkelling give excellent opportunities to observe the natural behaviour of fishes as they interact with each other and encounter other animals and plants. Unlike most wildlife on land, they are highly approachable. It is possible to get as close as if you were peering at a fish tank. However, this requires care and patience. Fishes need time to accept an unfamiliar presence and are generally wary at first but after 10, or perhaps 20, minutes they will usually start to relax, providing you have made no threatening moves in the meantime. It is best to keep almost still, and certainly to avoid sudden movement. Learn to recognize the signs that a fish is becoming agitated – a twitch usually indicates the intention to swim away, a raised dorsal fin is a message to back off. If the fish starts to

LEFT **The coral grouper** *Cephalopholis miniata* **is an approachable reef fish.**

retreat a short distance, it is best to do the same and let it set its own boundaries of personal space; going after it will only drive it speedily away.

Feeding fishes used to be popular with divers, but it sometimes works too well and 'tamed' fishes can become a boisterous nuisance or even a danger if they are large. Other dangers apply to the fishes, the loss of their natural caution leaves them vulnerable to spearfishing, and an unnatural diet often upsets their digestion.

LEFT **Although coral reefs are the primary home of coral fishes, shipwrecks and other manmade structures are colonized by them also, in suitable conditions. Here yellowsaddle goatfish** *Parupeneus cyclostomus* **swim alongside the coral-encrusted wreck of the** *Thistlegorm* **in the Red Sea, Egypt.**

Butterflyfishes and angelfishes

With their conspicuous colouring and habit of hovering daintily around the corals, butterflyfishes (family Chaetodontidae) are often popularly regarded as the most typical fish of coral reefs, and they are sometimes termed coralfishes. Angelfishes (family Pomacanthidae) are similar but some are larger and more spectacular. Both families are found on coral reefs throughout the tropics and they are much sought after by aquarists.

These fish require careful maintenance, however, and those species that feed on corals or sponges are particularly difficult to keep successfully.

Butterflyfishes may reach 30 cm (12 in) or more in length, as in the lined butterflyfish *Chaetodon lineolatus* (Indo-West Pacific), but most species are commonly seen at around half that size or less. The largest Indo-Pacific angelfishes, such the yellowbar angelfish

BELOW **Striped butterflyfish *Chaetodon fasciatus* are eye-catching on a coral reef.**

Pomacanthus maculosus, a magnificent blue-and-yellow fish familiar to divers in the Red Sea and Gulf of Oman, reach a maximum size of about 45 cm (18 in). Even larger species occur in the western Atlantic: the gray angelfish *Pomacanthus arcuatus* can be 60 cm (24 in) in length. At the other end of the scale are the pygmy angels of the genus *Centropyge*, of which the African pygmy angelfish *Centropyge acanthops* (western Indian Ocean) and the damsel angelfish *Centropyge flavicauda* (West Pacific) are no more than 6 cm (2 in) in length.

Similarities and differences

Because they share several characteristics, butterflyfishes and angelfishes were regarded as one family until recently. But although they are closely related there are significant differences between them, and they are now treated as two distinct families. The family name of butterflyfishes, Chaetodontidae, refers to the small bristle-like teeth (from the Greek words *chaite*, meaning hair, and *odontos*, tooth), a feature of the angelfishes too. Both have small mouths and deep bodies that are compressed laterally. Butterflyfishes have a more pointed snout, and they have particularly stout spines of the dorsal and anal fins. The most important characteristic that distinguishes the angelfishes is the strong, sharp spine projecting backwards from the preopercle (the anterior bone of the gill cover); butterflyfishes do not have this spine. This thorn-like feature is referred to in the angelfishes' family name of Pomacanthidae (from the Greek words *poma*, a lid or cover, and *acantha*, a thorn or spine).

RIGHT **The prominent spine on the gill cover of the yellowbar angelfish *Pomacanthus maculosus* is characteristic of the angelfish family.**

Where are they found?

The butterflyfish family is a large one, comprising about 120 species, whereas there are only about 80 species of angelfish. Both families are found on coral reefs throughout the tropics, but the majority of species live in the Indo-Pacific. Australia has the greatest diversity, with at least 50 butterflyfish species and more than 20 angelfish species. Some species, such as the threadfin butterflyfish *Chaetodon auriga* and the emperor angelfish *Pomacanthus imperator*, are widespread in the Indo-Pacific.

By contrast, a few others are endemic to one small region. For example the Easter Island butterflyfish *Chaetodon litus* occurs only around Easter Island, and the resplendent pygmy angelfish *Centropyge resplendens* is restricted to Ascension Island. The Red Sea, which is noted for its many endemic species, has seven butterflyfishes and two angelfishes that occur nowhere else. Conspicuous among these are two beautiful species: the striped butterflyfish *Chaetodon fasciatus* and the masked butterflyfish *Chaetodon semilarvatus*, both of which are easily approached by divers. The Caribbean Sea has far fewer butterflyfishes and angelfishes – only seven species of each – than the Indo-Pacific.

RIGHT **Emperor angelfish** *Pomacanthus imperator*.

LEFT **Gray angelfish** *Pomacanthus arcuatus.*

BELOW LEFT **Masked butterflyfish** *Chaetodon semilarvatus* **are endemic to the Red Sea.**

Butterflyfish groupings

The vast majority of butterflyfish species have been placed in one large genus, *Chaetodon*. They generally have a rather compact, oval to rectangular profile (excluding snout), unlike members of a smaller grouping – the genus *Heniochus*. Members of the latter group are known as bannerfish on account of the extended fourth spine of the dorsal fin, an eye-catching white pennant held high above the fish's back. The threadfin butterflyfish and a few other species of *Chaetodon* also have an extended fin, but in these it is just the soft rays (fourth to sixth) of the fin that form a fine, much less conspicuous, filament.

ABOVE **The compact shape of the striped butterflyfish is typical of the genus *Chaetodon*.**

ABOVE **A long pennant is typical of bannerfishes such as the Red Sea bannerfish *Heniochus intermedius*.**

Out and about on the reef

Butterflyfishes and angelfishes are active by day, seeking shelter at night between or beneath the branching corals, where their laterally compressed bodies allow them to pass through narrow gaps. They are generally confined to the shallower reefs, above 30 m (98 ft). A few occur in deeper waters, notably the modest or brown-banded butterflyfish *Chaetodon modestus*, found down to about 200 m (656 ft) in Hawaii and a few other localities in the Indo-Pacific, and the threeband butterflyfish *Chaetodon guyanensis*, which is found at similar depths in the Caribbean.

In most species of these two families the fish are solitary or live in pairs, with butterflyfishes (and perhaps also some angelfishes) often bonding for life. The graceful swimming motion of a pair is a delight to watch, as they move in synchrony, staying close together, facing the same direction and turning at the same moment. Butterflyfishes tend to stay in one general area, but their home patch is large enough to give them plenty of range for foraging. A few species, including the triangular butterflyfish *Chaetodon baronessa* (West Pacific), and others that feed exclusively on coral polyps, have a much more restricted range and are fiercely territorial in defending it. Other butterflyfishes are not much inclined towards territorial fighting, except for the odd spat in the scramble for night-time roosts. Many angelfishes are territorial, however. In pygmy and *Genicanthus* angelfishes, the dominant male maintains a harem of up to four females. If the male is removed, the most dominant female changes sex to replace him, in the course of only a few weeks. This phenomenon is well known also in basslets (family Serranidae) (see p. 44).

When butterflyfishes form schools they may travel further afield. Some species can be found schooling at one site but not at another, but only a few species school habitually, such as the pyramid butterflyfish *Hemitaurichthys polylepis* (West Pacific), and the bannerfish *Heniochus diphreutes* (Indo-Pacific). Whereas most butterflyfishes and angelfishes stay fairly close to the reef surface in order to find food, these fishes hover in midwater, where they feed on plankton.

How do they feed?

Butterflyfishes are well adapted to the coral reef environment, feeding on worms and other tiny invertebrates that live amongst the corals, and on the coral polyps themselves. These fishes have a pointed snout, which is often long and ends in a small mouth, enabling them to pluck these morsels from a stony thicket. This is taken to extremes in the longnose butterflyfish *Forcipiger longirostris*, an Indo-

Pacific fish with such an elongated snout that it can nip the soft tissues of tube feet and pedicellariae (minute pincer-like structures) out from between the spines of sea urchins.

Angelfishes will also eat some small active invertebrates, but encrusting animals and plants are their main diet. The larger angelfishes (*Pomacanthus* spp. and *Holacanthus* spp.) have a preference for sponges, and also eat sea squirts and sea

ABOVE **Schooling bannerfish *Heniochus diphreutes* feed on plankton in open water.**

TOP **An adult French angelfish** *Pomacanthus paru*.

BOTTOM **The colour pattern of juvenile French angelfish is spectacularly different from that of the adult.**

Colours

Butterflyfishes typically have yellow in their colour patterns, often combined with white, orange or black markings. Vertical or diagonal stripes are common, particularly a broad black band through the eyes. Their colour patterns may play a role in courtship, but are likely to be more important for defence. Angelfishes are varied, although their colours often include yellow and blue. Sexual differences in colour pattern do not occur in general but there are a few exceptions; some of the West Pacific angelfishes of the genus *Genicanthus* are strikingly sexually dimorphic, usually with the male strongly striped and the female plainer. However, there may be considerable differences between adults and juveniles.

Juveniles – colour and habit

Juvenile butterflyfishes may differ from adults in colour pattern, and are more likely to have a prominent false eye-spot towards the tail, but the differences are much more startling in some angelfishes. *Pomacanthus* juveniles, with contrasting bands of black and yellow or dark blue and white, often forming concentric rings, look nothing like the mature fish. In *Holacanthus*, too, there are marked disparities between young fish and adults; juveniles of several species have vertical blue stripes that disappear when the fish attains maturity. Juveniles of both genera are much more territorial than the adults. Some, particularly young gray angelfish *Pomacanthus arcuatus*, are likely to be seen in the sheltered lagoons behind the reef, while others may live on the reef proper. Several Caribbean species act as cleaners while young, picking ectoparasites off larger fishes. Young French angelfish *Pomacanthus paru* do this some of the time and gray angelfish occasionally. Their bold stripes may advertise their behaviour to their clients, and French angelfish have been observed to swim with a fluttering motion in the manner of certain habitual cleaners in other families.

Damselfishes and anemonefishes

Damselfishes (family Pomacentridae) are small, in the main less than 15 cm (6 in), but they are very abundant and diverse, with well over 300 species. They are found throughout the tropics, but mostly on Indo-Pacific reefs, and with only 16 species in the Caribbean. They generally stay in shallow waters, above 20 m (66 ft) depth, but several species, most of them pullers of the genus *Chromis*, have been found below 60 m (197 ft). Some damselfishes graze algae from the reef, while many others congregate in shoals to feed on minute planktonic animals in midwater, or are omnivores. The range of prey or other food items is limited only by what they can manage to bite with their small mouths.

Included in the damselfish family are the highly colourful anemonefishes (also known as clownfishes). They are classified as the subfamily Amphiprioninae and, although there are only a small number of species, they have captured popular imagination because of their associations with sea anemones.

BELOW **Plankton-feeding pullers of the genus *Chromis* hover above staghorn coral.**

Damselfishes' colours

Juvenile damselfishes are often exquisitely coloured and may be very different from the adults. As adults, they range from dull brown or grey to vivid blue, or are yellow, such as the golden sergeant *Amblyglyphidodon aureus*, a Pacific species that hovers alongside the walls of sea mounts and reefs. Most eye-catching of all members of the damselfish family are the clown or anemonefishes, which have orange and yellow hues emblazoned with broad white bands.

ABOVE **Golden sergeant *Amblyglyphidodon aureus*.**

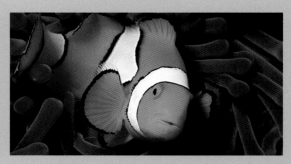

ABOVE **Bold colours of the clown anemonefish *Amphiprion percula*.**

Shoaling damselfishes

On a shallow reef with a healthy covering of staghorn coral, the layer of water immediately above the corals may appear cloudy from a distance. As you swim closer, the shimmering clouds will turn out to be shoals of small fishes, among them many damselfishes, such as turquoise or silver-grey pullers of the genus *Chromis*, each fish about 5 cm (2 in) in length. Their behaviour is interesting to watch, particularly when you can observe groups of pullers, and humbugs of the genus *Dascyllus*, above an isolated table coral. As you approach, the fishes retreat into the safety of the coral thicket in unison, as if pulled by invisible strings. Providing you remain still for a minute, they will rise out into the open again to resume feeding on plankton.

Shoaling damselfishes include various black-and-white striped sergeants of the genus *Abudefduf*. These are common all over the tropics at sites near the surface, above and

ABOVE **Scissortail sergeants *Abudefduf sexfasciatus* skim the top of a surf-swept reef.**

around coral reefs, some of which are exposed to surf or current, and where both plankton and algae are to be found. Late in the afternoon increasing numbers of them are sometimes noticeable, milling about right at the surface. They can be fierce little fishes, pecking at divers' hands or heads.

Territorial behaviour and breeding

The majority of damselfishes are territorial, particularly those that feed on the reef bed. Algal-feeders such as the gregories *Stegastes* (a genus with several Caribbean as well Indo-Pacific representatives) guard their patches jealously, thus 'farming' a thick growth of turf. In all cases, territorial behaviour increases to a feverish pitch during breeding. A courting couple will perform complex displays, including swimming together excitedly and chasing each other. The male may swim in an up-and-down pattern that has been likened by observers to a roller-coaster. These fish sometimes exhibit dramatic, although short-term, colour change before spawning at the chosen nest site, which has been plucked and cleaned in readiness. The male then assumes the role of protecting the eggs from predation by fishes that are often much larger than itself. Although the eggs are typically laid on the sea bed, various substrates may serve as nest sites. A dead

LEFT **The yellowtail damselfish *Microspathodon chrysurus* feeds on algae as an adult. This common Caribbean fish is territorial and bold, but not as aggressive as some other damselfish species.**

whip coral, for example, takes on a new appearance when a whitebelly damselfish *Amblyglyphidodon leucogaster* applies a coating of tightly packed eggs to the wiry stem. The fish watches over its eggs, pecking at them from time to time, probably to weed out dead ones. The larvae of nearly all damselfishes drift in open water before reaching a suitable site on the reef, where they can develop into adults.

Anemonefishes or clownfishes

If you swim over the top or upper slopes of an Indo-Pacific coral reef, you will see the varied shapes and colours of coral mounds and bushy outcrops spread below. Dotted amongst these, large sea anemones catch the eye as splashes of colour, of either the anemones themselves or of the small, brightly coloured fishes that live in close association with them.

ABOVE **A whitebelly damselfish** *Amblyglyphidodon leucogaster* **guards the minute eggs that have been attached to a whip coral.**

OPPOSITE **Two-band anemonefish** *Amphiprion bicinctus* **hover around a magnificent sea anemone** *Heteractis magnifica*.

An anemone will often have several fish living in association with it. The largest fish in the group is always female, the next is a mature male, and smaller fishes are immature males. If the matriarch dies, the largest male changes sex to replace her, and the next male in line becomes sexually mature. Thus the colony can continue to function for breeding. This type of sex reversal is a notable feature of anemonefishes – where sex reversal occurs in other reef fishes, it is more often in the other direction, from female to male.

Anemonefishes (or clownfishes) are captivating to watch. They never stray far from their sea anemone but they rarely seem to stay still either, as they hover over the tentacles or wriggle down amongst them. Sometimes a fish will disappear beneath the tentacles, and just as you are wondering where it has got to, the tentacles ripple and the fish pokes its head out, beady eye gleaming. It is impossible to guess exactly where the fish will pop up next. On an exposed site that is subject to currents, as is common in the Maldive Islands for instance, anemonefishes may have to swim hard at times to hold their position above the anemone. The fish waggles its pectoral and tail fins vigorously as it faces into the current, while the anemone's tentacles stream out below.

Papua New Guinea is a particularly good place to see anemonefishes; nine species live there and some of them are very abundant. At certain sites anemones and their attendant fish are everywhere you turn to look on the reef top and upper slopes. Anemonefishes are widespread in the Indo-Pacific region, but none is found in the Atlantic Ocean.

TOP **Clark's anemonefish**
Amphiprion clarkii.

BOTTOM **Another colour
variety of Clark's
anemonefish.**

Colourful clowns

Some species, such as the pink anemonefish *Amphiprion perideraion*, vary little in colour, but others have strikingly different colour varieties. Clark's anemonefish *Amphiprion clarkii* is perhaps the most variable. It has either two or three white bands, and its overall colour ranges from orange-yellow with little or no black, to a melanistic, almost completely black, form. As in other damselfishes, juvenile anemonefishes often change their colour pattern as they mature into adults.

Partners

Twenty-eight species of anemonefishes are known, along with 10 species of anemones that act as hosts. Some species of anemonefishes can live with several different anemone species, whereas others are restricted to one particular host. The least discriminating is Clark's anemonefish – it can use all 10 anemone species, and is the most widespread anemonefish. At the other extreme are several anemonefishes that live with one host only. These include the Seychelles anemonefish *Amphiprion fuscocaudatus*, which lives with Merten's sea anemone *Stichodactyla mertensii*, and the spine-cheek anemonefish *Premnas biaculeatus*, which lives with the bulb-tentacle sea anemone *Entacmaea quadricolor*. The Seychelles anemonefish occurs only around the Seychelles Islands and Aldabra, but the spine-cheek anemonefish has a much wider and more easterly range, from the Malay Peninsula to Melanesia. It is common around Papua New Guinea, where it may be seen nestling in anemones that have almost

The small number of species (only 28 in all) seems surprising, because they are such a familiar sight to divers on Indo-Pacific reefs. Their attractive appearance and behaviour endear them to aquarists, and most species have been heavily fished for this trade.

BELOW **The spine-cheek anemonefish *Premnas biaculeatus* lives exclusively with the bulb-tentacle sea anemone *Entacmaea quadricolor*.**

pure-white tentacles as well as in the more typical varieties, which are dull olive-green. The tips of the tentacles are characteristically bulbous and bob like balloons as the fish moves across the anemone.

Bold anemonefishes

Anemonefishes are surprisingly bold for their size, some more so than others. Saddleback or panda clownfish *Amphiprion polymnus* think nothing of giving a diver's bare hand a painful

Living with the enemy

Many large sea anemones are quite capable of living without anemonefishes, but the fishes are so dependent on the protection of their hosts' stinging tentacles that they are never found elsewhere in their natural environment, although they can be kept in an aquarium without an anemone. The benefits of the partnership are not one-sided though. The fish drive away butterflyfishes that are inclined to eat an anemone's tentacles, and they may clean away debris and parasites from the anemone.

The stinging cells (nematocysts) of the anemone are used for defence and to stun prey. So how do the fish avoid getting stung? The answer is that they are protected by a coating of mucus. A substance in the mucus coating of the fish acts in a similar manner to the anemone's own protection against stinging itself – in each case the chemicals inhibit the firing of the stinging cells. A popular theory is that the fish acquires the mucus from the anemone as its body brushes against the tentacles. An alternative suggestion is that the immunity is an adaptation of the fish's own mucus. Either way, investigations have shown that the fish needs to interact frequently with its anemone, and can be stung after a long separation.

Anemonefishes are not the only fishes associated with sea anemones: young dominoes or three-spot dascyllus *Dascyllus trimaculatus* often live with anemones. They are also damselfishes but are not classed as anemonefishes: instead they belong to the subfamily of pullers, the Chrominae. Other animals may share the fishes' living quarters: they include shrimps, particularly small, semi-transparent members of the large genus *Periclimenes*, as well as porcelain crabs.

FAR LEFT **Dominoes *Dascyllus trimaculatus* share a magnificent sea anemone with a two-band anemonefish *Amphiprion bicinctus*.**

LEFT **Anemonefishes (such as these pink anemonefish *Amphiprion perideraion*) are not stung by the tentacles of the anemone.**

LEFT **The skunk anemonefish** *Amphiprion akallopisos* **seen here nestling against a magnificent sea anemone has a large fishlouse attached to it. This parasite is the isopod** *Renocilia heterozota* **(a member of the Crustacea).**

nip, and repeating the attack until the intruder is forced to beat a hasty retreat. Breeding male fish are the most aggressive, and never more so than when they are guarding their eggs. The eggs are not difficult to spot because they are bright orange or red, and laid in a dense patch on a rock or other hard surface beside the anemone. As with most damselfishes, the male tends the eggs until they hatch, fanning them with his fins and removing debris. Saddleback clownfish often live with Haddon's sea anemone *Stichodactyla haddoni*, which protrudes from sandy patches. The anemone often has other occupants as well, such as porcelain crabs, but these are timid and leave the heroics to the cohabiting fish. Haddon's sea anemone has its own means of defence against unwelcome intruders. It shrinks away from contact with anything other than its normal occupants, and may even disappear completely beneath the sand, leaving its anemonefish to mill about over the spot, anxiously waiting for it to reappear.

BELOW **A saddleback anemonefish (panda clownfish)** *Amphiprion polymnus* **guards eggs, which have been laid alongside Haddon's sea anemone** *Stichodactyla haddoni*.

Groupers and seabasses

BELOW **A jutting lower jaw is typical of groupers, such as the red hind *Epinephelus guttatus*. The tiny fish on the grouper's cheek is a cleaning goby of the genus E*lacatinus*.**

Groupers and seabasses (family Serranidae) are predators of smaller fishes and crustaceans, but rather than waste their energy in darting about after their prey, they spend much of the day lying in solitary wait. Most members of the family are heavy-bodied, and they can often be found resting in the shelter of coral heads on the reef. Their lair might be beneath an overhang or spreading branches of corals; they may hover there, just off the sea bed, or perch outside on a nearby rock.

When suitable prey passes by, they make a quick lunge and dispatch it efficiently by expanding their large mouth to engulf it. Their jaws are equipped with bands of small sharp teeth, and usually a few larger canines at the front. The lower jaw often juts out, giving the fish a rather grim air, and the lips are generally large.

ABOVE **A blacktip grouper** *Epinephelus fasciatus* **perching on coral rocks.**

ABOVE RIGHT **Nassau grouper** *Epinephelus striatus*.

Groupers tend to be slow to move away and many species will allow divers to approach them closely. Large groupers are an attraction on certain sites, where they hang around divers, either hoping to be fed titbits or merely out of curiosity. Individuals of one species, known as the potato cod *Epinephelus tukula,* have been hand-fed by divers, but a fish that may be up to 2 m (6.5 ft) in length and may weigh over 100 kg (220 lb) can be dangerous if it becomes too eager! A Caribbean counterpart, the Nassau grouper *Epinephelus striatus*, may be about half that size. However, many groupers, the Nassau grouper among them, are foodfishes of commercial importance and larger individuals are uncommon on heavily fished reefs.

In most grouper species all the fish are female to start with and some change sex to become breeding males. They commonly spawn in pairs but a few species, including the Nassau grouper, form large aggregations for spawning.

Grouper and seabass diversity

Worldwide there are about 500 species of groupers and seabasses, the majority of which are coral-reef dwellers. They include wide-ranging species such as the giant grouper *Epinephelus lanceolatus*, which, although rarely seen, occurs from the Red Sea eastwards to the Pitcairn Islands. Other species are much more restricted, such as the Tonga grouper *Epinephelus chlorocephalus*, which is endemic to Tonga. Most members of the family live at comparatively shallow depths, where divers often encounter them, but some are deep-water species. The golden grouper *Cephalopholis aurantia*, for example, usually lives below 50 m (164 ft) and sometimes even below 200 m (656 ft), while some Caribbean basses of the genus *Serranus* range down to around 400 m (1310 ft). Although the typical mode of life for this family of fishes involves keeping a low profile on the reef, some are seen in more open territory, for example lyretail groupers of the

genus *Variola*, and the basslets (subfamily Anthiinae). Lyretails and basslets have a forked or concave tail margin, which is suited for these active swimmers but unusual in other members of the family.

Fishes in this family are are very diverse in size and shape, and some of them, such as the small basslets and the hamlets, are very different both in appearance and behaviour from the norm. In contrast to these are some spectacular heavyweights.

Huge groupers

A few groupers are enormous but rarely seen. The giant grouper *Epinephelus lanceolatus* of the Indo-Pacific grows to about 3 m (10 ft) and weighs up to 300–400 kg (660–880 lb), making it the largest bony fish found on coral reefs. For all its monstrous appearance it is timid and may be glimpsed occasionally hiding well back in dark caves. The western Atlantic has a similar giant of impressive bulk, the jewfish *Epinephelus itajara*, reaching well over 2 m (6.5 ft) and weighing over 300 kg (660 lb). The jewfish also uses a cave

or a wreck as its refuge, where it can rest while digesting a hawksbill turtle or some other substantial meal. Both fish species have suffered depletion from spearfishing, the jewfish particularly in Florida, but when given the chance they may live for 35–40 years.

Colour patterns

Many groupers are patterned over most of the body with spots or blotches. In the larger species (often reaching a length of 50 cm [20 in] or more), neutral colours such as dark brown and cream predominate, and their dappled bodies blend in with the surroundings. A few large groupers are more colourful. The tomato grouper *Cephalopholis sonnerati* may be bright red as its name suggests, but Pacific members of the species can change rapidly to brown shades. Lyretail groupers of the genus *Variola*, widespread in the Indo-Pacific, are sometimes red or lavender, with blue or purple spots. In the Caribbean, the tiger grouper *Mycteroperca tigris* (occasionally growing to as much as 1 m [39 in]) may assume a red hue over its stripes.

ABOVE LEFT **The lyretail grouper *Variola louti* is one of the more active groupers on the reef.**

BELOW **The mottled pattern of the camouflage grouper *Epinephelus polyphekadion* disguises the fish as it rests on the reef.**

ABOVE **White-lined grouper or rockcod *Anyperodon leucogrammicus*.**

By any other name?

The bewildering variety of different common names applied to the same species or group of fishes is particularly noticeable with the grouper and seabass family. In Australia, where many groupers (subfamily Epinephelinae) occur, they are known as gropers, and some are referred to as cods, rock-cods or coral trouts, although they are not related to the true cods or trouts. Caribbean members of the family are often called seabasses or basses (not to be confused with basses of temperate seas, which do not belong in this family); these terms are sometimes restricted to one of the subfamilies (Serraninae), whose members are fairly small and include the hamlets.

One of the three subfamilies of the grouper and seabass family, the Anthiinae, are called basslets,

fairy basslets or sea-perches. Confusingly, the name basslet is also used for other fishes: a small tribe of groupers, the Liopropomini (secretive, mainly deep-water fishes), some of the smaller seabasses of the subfamily Serraninae, and a separate family (the Grammatidae), as well as for certain members of the dottyback family (Pseudochromidae) and the prettyfin family (Plesiopidae).

Soapfishes were formerly regarded as a separate family (the Grammistidae) but are now included in the grouper subfamily Epinephelinae. Many soapfishes produce a bitter tasting toxin – grammistin – from their skin, which acts as a deterrent in their natural environment and can kill other animals in the confines of an aquarium.

ABOVE **Tiger grouper**
Mycteroperca tigris.

RIGHT **Red colour phase
of the coney**
Cephalopholis fulva.

The brightest colours are generally reserved for the medium-sized groupers. In many of these, bright blue spots with a dark surrounding ring stand out vividly, giving rise to names such as jewelspotted and jewel grouper (alternative names for the peacock grouper *Cephalopholis argus* and the coral grouper *Cephalopholis miniata*, respectively). The coral grouper is one of several red species and is a common fish on Indo-Pacific reefs. A Caribbean counterpart, the coney *Cephalopholis fulva*, is also red in one of its colour phases, but it is very variable, sometimes bicoloured or occasionally brilliant yellow. The blacksaddled coralgrouper *Plectropomus laevis* (sometimes known as the Chinese footballer, or footballer trout) is another with different colour phases. The juveniles, and sometimes also adults, are boldly banded and are thought to mimic the poisonous pufferfish *Canthigaster valentini*, although the resemblance is less marked than that of the pufferfish's other mimic, the blacksaddle filefish (see p. 63).

Many larger groupers are able to change colour according to their surroundings, particularly if they need to avoid being seen. On such occasions they may turn darker or paler, and their markings meanwhile may become indistinct or, alternatively, form a series of vertical bands that break up the fish's outline. Thus they can be highly visible one moment, for social interactions, and escape notice the next, as stealthy predators.

Hermaphrodite hamlets

Among the smaller serranids are the hamlets, a Caribbean genus (*Hypoplectrus*) of fishes with amazing mating behaviour. They are hermaphrodites whose sexual organ produces both sperm and eggs simultaneously. When the fish pair off at dusk to spawn, one partner takes on the female role and the other acts as the male, but they have been observed to alternate during the evening session. The pair rise above the sea bed, wrap their bodies around each other and vibrate as they slowly sink again, repeating this brief sequence many times. During courtship the fish make sounds, although these are generally inaudible to the human ear.

All the species have similar courtship rituals and they appear to recognize their own kind solely by colour pattern. Each species of hamlet is distinctive in colour but some of them resemble other types of fishes, particularly damselfishes. Hamlets and damselfishes may be much the same size, around 10 cm (4 in) in length, and both are very territorial, but whereas damselfishes feed on algae, hamlets are carnivorous like most other members of the family Serranidae.

Predatory hamlets may gain an advantage in resembling harmless herbivores when they are seeking prey, and the crustaceans they prey on probably do not have sharp enough vision to detect the finer points of difference.

ABOVE **Brightly coloured coral grouper** *Cephalopholis miniata*.

LEFT **Barred hamlet** *Hypoplectrus puella*.

One example of this resemblance (whether it has arisen through mimicry or coincidence has yet to be established) is that of the yellowtail hamlet *Hypoplectrus chlorurus* to the yellowtail damselfish *Microspathodon chrysurus*, both dark-bodied fishes with yellow tails.

Tiny basslets

At first sight the basslets or anthiases, of the subfamily Anthiinae that are commonly seen on coral reefs, seem to have little in common with other members of the family Serranidae. They are mostly small, often slender, and the tail is generally lunate (i.e. with the margin curved as in a new moon), not truncate (squared off) or rounded as in the majority of groupers. The males often have filaments extending from the dorsal fin or the tail, which they display by erecting the fin. Basslets are mostly orange or magenta, and are sometimes referred to as 'goldfish' because of their superficial resemblance in size, shape and colour to those unrelated freshwater fishes.

Unlike typical lone groupers, basslets gather in large schools out in the open, feeding on plankton (including the eggs of other fishes) above the reef. In tranquil situations, basslets dart about in all directions as their keen eyes spot minute morsels to prey on, but they never stray far from their patch of reef – often on the edge of a reef wall or drop-off. When a threat looms or if there are strong currents they become a unified shoal. They then move as if choreographed, in the one scenario making a prudent bolt, in the other, fluttering their pectoral fins to stay swimming on the spot. Male basslets have a harem of

females, the dominant one of which will usually change sex to take his place if he dies.

Basslets are common and conspicuous on Indo-Pacific reefs, but only a few species occur on coral reefs in the Atlantic. Other members of the Anthiinae are bottom-dwelling, but they are mainly deep-water species. Similar small colourful fishes in the Caribbean are also known as basslets or fairy basslets but they are a distinct family, the Grammatidae. Unlike the Anthiinae, they are solitary. The best-known member of this small group is the fairy basslet *Gramma loreto*, no more than 8 cm (3 in) in length but remarkable for its bicoloured beauty of purple and yellow.

ABOVE **Basslets of the subfamily Anthiinae: scalefin anthias** *Pseudanthias squamipinnis*.

Wrasses

Wrasses are a large and immensely varied family, the Labridae, with a wide range of sizes, shapes, colours and behaviour. It is hard to picture a massive 2 m (6.5 ft) humphead wrasse as a relative of tiny species that are no more than 5–6 cm (about 2 in) in length, such as the minute wrasse *Minilabrus striatus*, of the Red Sea, and the pelvic-spot wrasse *Pseudocheilinops ataenia*, found mainly in Indonesia and the Philippines. The smallest Caribbean representative, the dwarf wrasse *Doratonotus megalepis,* is only slightly larger than those two. Although many wrasses are slender-bodied – the cigar wrasse *Cheilio inermis* is an aptly named example – some are heavily built. The humphead wrasse is the most extreme example but tuskfishes of the Indo-Pacific genus *Choerodon* may also be sturdy, and the Caribbean hogfish *Lachnolaimus maximus* is a deep-bodied species. These and other sizable wrasses are fished for food.

BELOW **Wrasses, such this member of the genus** *Cheilinus*, **often have complex colour patterns.**

Where are they found?

About 500 wrasse species occur worldwide. The majority live on coral reefs, but others live in temperate seas and some even survive in cold sub-Arctic waters. They are found in every part of the reef environment, although they tend to be more prevalent at shallower sites. Some of the most secretive wrasses live in caves and are rarely seen, while others are conspicuous out in the open. Wrasses are generally seen singly, or in pairs or scattered groups. A few, however, may gather in larger numbers, particularly plankton-feeders, such as the flashers of the genus *Paracheilinus* and their relatives in the Indo-Pacific, and the creole wrasse *Clepticus parrae*, which is abundant on outer reefs of the Caribbean.

Wrasses retire early for their night's rest, either retreating into crevices or burrowing into the sand, *Coris* wrasses and some of the smaller species taking the latter mode. Razorfishes (*Xyrichtys* and related genera) take

ABOVE **Humphead wrasse *Cheilinus undulatus*.**

ABOVE **The humphead wrasse's domed forehead becomes more prominent as the fish matures.**

Giant wrasses

The largest of all wrasses, the humphead wrasse *Cheilinus undulatus* of the Indo-Pacific, may grow to over 2 m (6.5 ft) and weigh up to about 190 kg (418 lb). The species is well known to divers, and certain individual humpheads are a popular feature at some sites. In the Ras Mohammed National Park, Egypt, for instance, they are accustomed to divers and will allow close approach, sometimes tagging along behind a group. However, they are wary in places where spearfishing occurs, and fishing in general has made the species vulnerable. Apart from the impressive size, these fish have a distinctive appearance, the hump on the forehead and the thick lips swelling with maturity. Humpheads are generally seen along the outer reef slopes, where they search for formidable prey. They will even eat crown-of-thorns starfish, and can crush heavy turban shells and top shells.

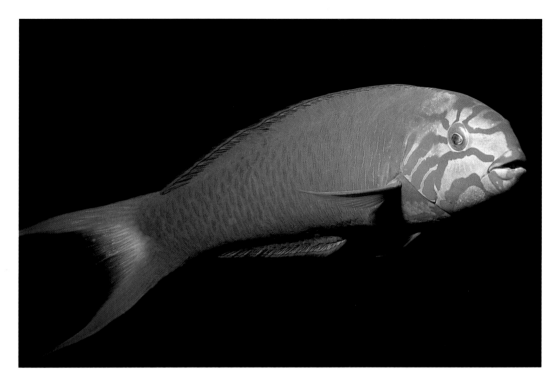

LEFT The vividly
patterned crescent
wrasse *Thalassoma
lunare*, and related
species, often live
on shallower parts
of the reef.

cover during the day also. They dive headlong into the sand, aided by their compressed shape and the keel on their forehead.

Breeding

Sex change from female to male takes place in most wrasse species, and in some genera – *Cirrhilabrus* and the cleaner wrasses *Labroides,* for example – it is common for a dominant male to have a harem of several females. The male courts a female by displaying to her with his fins spread. Spawning pairs, or sometimes groups, swim energetically towards the surface, release their eggs and sperm, then descend immediately. The activity is finished by the end of the afternoon or at dusk.

Active swimmers

Wrasses, like parrotfishes (family Scaridae), typically swim using just their pectoral fins in a rowing motion, and bring the tail into action only when they need to move rapidly. Although most wrasses frequent fairly calm water, the plankton-feeding species may swim in currents, and some others can cope with turbulent conditions and even prefer them. Among the more active swimmers are wrasses of the genus *Thalassoma* that inhabit exposed shallow zones on the outer reef, buffeted by waves and currents. One of these, the surge wrasse *Thalassoma purpureum*, escapes from predators by darting right into the extreme surge.

Changes with age and sex

In wrasses, the rites of passage from juvenile to adult and to terminal (i.e. breeding) male are commonly marked by gradual colour changes. Male wrasses tend to be more brightly coloured than the females, but the reverse is occasionally true, as in the feminine wrasse or blue-striped orange tamarin *Anampses femininus*. Some of the most notable changes take place in wrasses of the Indo-Pacific genus *Coris* and in hogfishes of the genus *Bodianus*. The clown coris *Coris aygula* is one of several species in which the conspicuous orange-and-white patterned juveniles bear little resemblance to their plainer elders. Certain other juveniles, again unlike the adult fishes, are camouflaged and mimic vegetation for protection.

Projecting fin filaments give young rockmover wrasses *Novaculichthys taeniourus* a frilly outline and resemblance to a clump of drifting seaweed (and also earn them the alternative names of dragon or reindeer wrasse). Similarly, indianfish (as juvenile blue razorfish *Xyrichtys pavo* are known) mimic a leaf swaying as it is carried by the current, and young dwarf wrasses are camouflaged amongst seagrass in the Caribbean.

RIGHT **Like many wrasses, Diana's hogfish *Bodianus diana* changes its colour pattern as it grows. The fish seen here is an adult – juveniles are dappled all over.**

LEFT **The hogfish *Lachnolaimus maximus* is a common Caribbean wrasse.**

Diverse feeding habits

Wrasses are closely related to parrotfishes. The second set of jaws, in the pharynx, is adapted for milling food in both these families, and in certain other fishes, but more strongly developed in parrotfishes than in wrasses. Unlike grazing parrotfishes, wrasses are carnivores, but within that definition their prey and means of catching it are as varied as the fishes themselves. Many take invertebrates from the sea bed but some feed on plankton and a few eat coral polyps, prey on small fishes, or are specialized in removing parasites from other fishes. Generally though, wrasses favour hard-shelled prey, particularly molluscs and crustaceans. The hogfish *Lachnolaimus maximus*, one of the larger wrasses, regularly dispatches crabs, hermit crabs, barnacles and sea urchins as well as a wide range of molluscs. Whereas many other fishes are unable to break up well-armoured animals, wrasses crush them with their pharyngeal mill. Like the herbivorous parrotfishes, they are able to eat on the move, a good strategy for avoiding unwanted attention in open territory. If the prey is too large to swallow whole the wrasse may pick it up and smash it against a rock.

Colourful names

The markings of wrasses display a spectrum of hues, and more than one species has justifiably been called 'rainbow wrasse'. Other names that reflect the colourful nature of their owners are the harlequin tuskfish *Choerodon fasciatus* (an orange-banded wrasse) and the psychedelic wrasse *Anampses chrysocephalus*. The flashers of the genus *Paracheilinus*, although no more than 10 cm (4 in) in length, are particularly exquisite. The males display by suddenly erecting their brilliant fins, which have long filaments in some species, and rapid colour changes may enhance the spectacle.

Many wrasses have narrow stripes on the face – salmon-pink, fluorescent green, electric blue or violet. Members of the genera *Cheilinus* and *Oxycheilinus* are often called maori-wrasses because of the supposed resemblance of their markings to Maori facial tattoos.

One of them, the cheeklined or bandcheek wrasse *Oxycheilinus digrammus*, has been known to effect a rapid colour change when swimming with goatfish, in order to blend in with the group, until it reveals its predatory nature by darting out from the group to snatch a small fish.

ABOVE **Cheeklined wrasse *Oxycheilinus digrammus*.**

ABOVE **Redbreasted wrasse *Cheilinus fasciatus*.**

Rock movers

The front teeth (canines) of wrasses are characteristically stout, often protruding like fangs or giving a bucktoothed appearance. Members of the genus *Choerodon* are called tuskfishes on account of their strongly developed canines, which, like some other wrasses, they use to shift coral rubble in their search for small invertebrate animals. They can grasp large chunks of coral in their teeth and turn them over. One of the razorfish wrasses, *Novaculichthys taeniourus*, is so adept at this that it has earned the name of rockmover wrasse.

Thicklips, tubelips, slingjaws and flashers

A noticeable feature of wrasses is their thick lips. These are particularly large and fleshy in the humphead wrasse *Cheilinus undulatus* and also in members of the genus *Hemigymnus*, which are commonly known as 'thicklips'. The lips may aid suction in some fishes that feed by hitting the sea bed and sucking in a mixture of sand, detritus and tiny invertebrate animals.

The diversity of wrasses has given rise to some bizarre structures. They typically have a rather protrusible mouth but this is taken to extremes by the slingjaw *Epibulus insidiator*. As it approaches potential prey, the fish distends its jaws in a sudden motion to form a tube half the length of its body, and sucks in the prey instantaneously. In bird wrasses of the genus *Gomphosus*, the mouth is at the end of a long narrow snout, which is used to probe crevices in the coral for small invertebrates. Another strategy is seen in tubelip wrasses of the genera *Labrichthys* and *Labropsis*. Their closed lips form a short tube that aids them in plucking coral polyps.

A highly unusual feature of a small group of wrasses is also believed to be a feeding aid. This is the double pupil of the flashers of the genus *Paracheilinus* and their relatives. It has been suggested that it acts as a magnifying lens, enabling the fishes to spot the minute planktonic animals that they prey on. The flashers differ from typical wrasses in having thin lips and a small mouth better suited to the plankton-feeding mode.

Cleaners

The specialized feeding activities of certain wrasses not only give them access to a plentiful, if unusual, source of food, but also provide a vital service. Cleaner wrasses (*Labroides dimidiatus* and a number of other species) pick parasites off other fishes and remove dead and diseased tissues from around wounds. Their territories, occupied by one or more individuals, are known as cleaning stations, conspicuous sites visited by fishes requiring treatment. A shallow Indo-Pacific reef will have a number of stations, where fishes of various shapes and sizes can be seen hovering as they await their turn to be cleaned by the small wrasse. Cleaner wrasses are very slender and generally less than 10 cm (4 in) in length, so they are able to attend to the gills and the inside of the mouth of larger customers. This would make them seem vulnerable but they are never eaten. A combination of boldly striped colour patterns and distinctive swimming routines ensures that cleaners are recognized and appreciated.

However, another less common fish of similar appearance takes advantage of this – the mimic blenny or false cleaner *Aspidontus taeniatus* poses as a cleaner but bites healthy skin instead.

A cleaning station is ideal for fishes in the vicinity and those that roam the reef, but what of others? Small territorial fishes are not prepared to leave their patch and instead need a cleaner to come to them. The wandering cleaner wrasse *Diproctacanthus xanthurus* is one of a few species that do just this, paying house calls on clients, chiefly damselfishes. Cleaner wrasses and the wandering cleaner wrasse are in the tubelip group of wrasses, the majority of which perform a cleaning service. This is a job for life as far as the *Labroides* cleaner wrasses are concerned; the others pursue it as juveniles but usually give it up on becoming adult, and turn to plucking coral polyps instead. Various other cleaners include the western king wrasse *Coris auricularis*, an Australian species in which juveniles and females set up cleaning stations, and juveniles of a few Caribbean species. Much of the parasite removal in the Caribbean is carried out by gobies, but there are some notable wrasse practitioners there also, such as the bluehead *Thalassoma bifasciatum*, a common reef fish, and the Spanish hogfish *Bodianus rufus*.

One of the most unusual practices is that of the twotone or cleaning sand wrasse *Thalassoma amblycephalum*. This agile swimmer eats tiny crustaceans in the plankton over much of its range, but in the Maldive Islands it has become a specialist feeder of the parasites on manta rays and other very large fishes.

Grazers and crushers

Fine wisps of seaweed grow over the tops of shallow reefs, covering dead corals and coral rock with a green matting. Various fishes and other animals crop this turf as their main source of food, and by doing so play a role in maintaining the healthy diversity of the reef community. Because their mode of grazing removes the weed filaments right down to the bare rock and below, it limits the spread of the algal mat and exposes bare surfaces. These are ideal attachment sites for invertebrate larvae, among them the early stages of reef-building corals. The cleared surfaces are also colonized by coralline algae, which are important in stabilizing areas of loose rubble by encrusting them and cementing the pieces together.

Parrotfishes and surgeonfishes

Two families of fishes are chiefly responsible for this grazing, parrotfishes (Scaridae) and surgeonfishes (Acanthuridae). Although their feeding activities may be beneficial in promoting diversity of coral-reef inhabitants, parrotfishes can be destructive. They are major producers of sand ground down from corals. As they move on from one patch of coral to the next some leave tell-tale signs of their feeding, white scars of newly exposed limestone showing the characteristic double groove made by the teeth, or 'beak' of certain species.

The two families have a similar number of species, about 70 to 80, and are found throughout the tropics, although relatively

LEFT **The princess parrotfish *Scarus taeniopterus* feeds by scraping algae from the coral surface.**

RIGHT **Redband parrotfish** *Sparisoma aurofrenatum*.

BELOW **Bumphead parrotfish *Bolbometopon muricatum* are often in schools.**

few surgeonfish species occur in the Caribbean. Some surgeonfishes and parrotfishes are solitary; others may gather in small groups or even vast shoals. Most fishes grow to around 20–50 cm (7–20 in) in length but a few reach an impressive 1 m (3 ft) or more. Among the largest are the rainbow parrotfish *Scarus guacamaia*, in the western Atlantic, and the bumphead parrotfish *Bolbometopon*

muricatum, in the Indian and Pacific oceans. Bumphead parrotfish have an appetite to match their massive bodies and their feeding has a great impact on the reefs they inhabit. Divers sometimes refer to them as buffalo fish, an apt name for a dense pack of heavy-bodied grey fish that move rapidly and keep close to the ground. The noise, as they all bite and scrape at the coral, grinding and crushing fragments, sounds like the rumble of a stampeding herd. Behind them drift clouds of sediment, expelled waste of coral sand. Parrotfishes can have a marked effect on the water clarity off a coral reef as the day wears on!

Surgeonfishes are typically oval in shape and more laterally compressed than parrotfishes. Some surgeonfishes are dull shades of brown or grey but others are colourful, with blue featuring strongly in some species, such as the powder-blue surgeonfish *Acanthurus leucosternon*.

Colour differences

Parrotfishes usually show strong differences in colour between the sexes and phases. The terminal phase consists of mature breeding males, which are typically brightly coloured, with greens and blues predominating. Males of the initial phase, together with females, tend to be muddy shades of brown, purple and pink. Initial-phase males spawn in groups, while a terminal male commands a female to itself. Terminal males are the outcome of sex change, along with colour change, of some of the females – a scenario similar to that in a closely related family, the wrasses.

Unlike parrotfishes, surgeonfishes do not show obvious sexual differences in colour but the juveniles may undergo marked change, sometimes in shape as well as in colour. A striking example relating to colour is seen as the blue tang *Acanthurus coeruleus* matures. This surgeonfish is common in the Caribbean

BELOW **Terminal phase colours of a male stoplight parrotfish** *Sparisoma viride*.

LEFT **A female or an initial-phase male of the stoplight parrotfish.**

and throughout the tropical west Atlantic, but it could easily be mistaken for three species instead of one. The young fish are brilliant yellow, as in a number of other species, and intermediates are sometimes seen with a blue body and yellow tail before they become entirely purplish-blue as adults. Two Indo-Pacific surgeonfishes, *Acanthurus tristis* and *A. pyroferus*, have juvenile colour phases that mimic certain angelfishes of the genus *Centropyge*.

Adapted for grazing

While parrotfishes and surgeonfishes have similar dietary requirements on a coral reef, they are not closely related and have quite different adaptations for grazing. Parrotfishes and surgeons eat only during the day. At the end of the afternoon they make their way back from their feeding grounds, and as night falls they seek out places to wedge themselves in amongst the corals. Shrouded by branches, little of the fish may be exposed, but the beam

of a diver's light may pick out a vivid patch of colour of a male parrotfish here and there on the reef. Later at night some parrotfishes surround themselves with a filmy mucus cocoon, possibly affording them protection against predators although this is unproven.

BELOW **The bucktooth parrotfish *Sparisoma radians* is a small species that browses on seagrasses and algae.**

LEFT **The steepheaded parrotfish** *Chlorurus gibbus* **has powerful jaws, which enable it to excavate coral.**

Parrotfishes

Parrotfishes have specialized dentition for highly efficient grazing. Their teeth are fused, often forming a beak-like structure, enabling them to scrape off the finest algal growth, and some species also excavate the surface layers of coral. The mix of algae and limestone is then ground down by their other set of teeth, on bones in the pharynx which are modified into a mill. Parrotfishes break down considerable quantities of dead coral in this way and a few bite live coral also. They have a very large intestine to cope with the digestion of nutrients from this unpromising material. Like their close relatives, the wrasses, they can snatch a meal and break it down afterwards, extracting the nutritious parts in a leisurely fashion. Not all parrotfishes feed by scraping or excavating; some browse on seagrasses and certain other sizeable plants that they are able to bite directly.

Surgeonfishes

Surgeonfishes, like parrotfishes, have a long intestine. Some species, mainly those that ingest sand with their food, have a gizzard-like stomach to aid digestion. The mouth is small, however, and instead of a parrot-like

BELOW **A yellow or white sheath, on each side of the tail base, covers the scalpel-like blade of the blue tang** *Acanthurus coeruleus.*

beak they have small, close-set teeth for plucking and nibbling. Many surgeonfishes are territorial in defense of their grazing grounds, but they employ a range of tactics for feeding, varying even within the same species. Surgeonfishes often form large aggregations when feeding, sometimes involving a mixture of species. They can be seen moving purposefully along the crest of the reef and on to the shallow flats in search of good grazing. As soon as a few swoop down and start plucking at the wispy algal growth, the rest immediately descend on the same patch, heads down and jostling each other in their haste to get at the morsels, as if scrapping over titbits thrown to them. Suddenly, as if by an unseen signal, they all move on a few metres to the next spot. The focus of their excitement may appear to be a dull stretch of ground but they find plenty to consume as they range the reef.

Not all members of the surgeonfish family consume algae, several species eat zooplankton instead, or as a supplement. This is common practice among the unicornfishes (Indo-Pacific surgeonfishes that have a frontal horn). They are large fishes, commonly reaching 50 cm (20 in) and occasionally 75 cm (30 in), and are often seen in large schools feeding in open water. The spotted unicornfish *Naso brevirostris*, and also the elongate surgeonfish *Acanthurus mata*, graze algae only while they are juveniles, progressing to plankton later. The few unicornfishes that have an adult preference for algae, including the widespread bluespine unicornfish *Naso unicornis*, eat the tough brown *Sargassum* weed, which is unpalatable to the majority of herbivores.

TOP **Bluespine unicornfish** *Naso unicornis*.

BOTTOM **Powder-blue surgeonfish** *Acanthurus leucosternon* **are grazers of algae.**

Surgeon's scalpel

Surgeonfishes owe their name to the blade-like plates, one on either side of the base of the tail in the sub-family Acanthurinae, that can cut as effectively as a scalpel. Another name for the fish, 'tang', meaning spike, refers to the same feature. The plates are small but made visible by their contrasting colour, and sometimes emphasized by a surrounding ring. In the blue tang, for example, they are yellow against the blue or purple body of the fish. The blades may be weakly venomous in a few species, such as the convict surgeonfish *Acanthurus triostegus*, which occurs in the Indo-Pacific.

Unicornfishes (subfamily Nasinae) usually have two pairs of blades, which develop a sharp keel as the fish matures and are fixed, whereas in surgeonfishes of the subfamily Acanthurinae they are retractable. In all cases the blades are used if necessary for defence against predators and in territorial clashes.

Other grazers

Surgeonfishes and parrotfishes often run up against herbivorous members of the damselfishes, which protect their own exclusive patches of algae. Damselfishes are highly aggressive and fully capable of driving away individual competitors much larger than themselves, but when surgeonfishes or parrotfishes shoal they can overrun a damselfish's territory by sheer force of numbers. Certain damselfishes get their own back: for example the gregories, *Stegastes* species, in particular, gang up with others of their species to guard their patch more successfully.

Parrotfishes and surgeonfishes also compete with other herbivorous fishes: pygmy angelfishes and combtooth blennies (see p. 26 and p. 75 respectively), and rabbitfishes (family Siganidae). Rabbitfishes are related to surgeonfishes and graze algae in a similar manner to them, usually on shallow reefs. Many of them are speckled or blotched and less conspicuous than surgeonfishes, but they are well armed, not with blades as in surgeonfishes, but with venomous fin spines, capable of giving a painful sting to unwary fishermen among others. Not all herbivores are fishes – sea urchins may have a significant impact on and around coral reefs.

Puffers and allied crushers

Puffers and their relatives have wide-ranging diets. Sometimes they graze on algae and seagrass in addition to eating various invertebrate animals that are either slow-moving or attached to the reef. Among those that include algae and seagrass in their choice of food are sharpnose puffers or tobies of the genus *Canthigaster*, as well as trunkfishes, filefishes and a few species of triggerfish such as the black durgon *Melichthys niger*. The durgon and others also consume plankton; but in general puffers and their allies are crushers and crunchers of food. They attack hard-shelled animals such as crabs and molluscs; spines of starfishes do not deter them either, and even the formidable crown-of-thorns starfish falls prey to the whitespotted puffer *Arothron hispidus*. Various puffers, and others including the titan triggerfish, bite the tips off coral branches. Filefishes are often daintier feeders, the pointed snout enabling those that eat coral to get at the polyps without breaking branches. The scrawled filefish *Aluterus scriptus* has been known to eat fire 'corals' of the genus *Millepora*, which are not true corals but hydroids. Hydroids are avoided by most animals (and divers) because of their stinging cells.

Puffers and their allies have small mouths but their dentition enables them to break hard items into bite-sized pieces. The teeth of puffers (family Tetraodontidae) and porcupinefishes (family Diodontidae) form a strong parrot-like beak, consisting of four and two plates, respectively. Those of other families in the group, including triggerfishes (Balistidae), filefishes (Monacanthidae) and trunkfishes (Ostraciidae), have few, but often large, teeth. They tend to be specialists in eating the animals that most other fishes find too well-protected to tackle, and some have an unusual method of getting under their prey's defences – a water jet. Triggerfishes blow water to roll sea urchins over, exposing the less spiny underside, and trunkfishes blow

OPPOSITE **Whitespotted puffer *Arothron hispidus*.**

sand away to uncover animals hiding there. Defending themselves calls for other characteristics.

Puffers and triggers – defence strategies

When faced with danger from a predator, slow-swimming puffers cannot make a quick getaway but they have another means of escaping attack – they inflate their bodies by ingesting water. This ability to make themselves appear larger and more formidable is shared with porcupinefishes, and fishes in both families have been given other names that reflect their unusual habit: puffers are sometimes called blowfishes, blowies or toadfishes, and one species of porcupinefish is known as the balloonfish.

Porcupinefishes have the added defence of being as spiny as their name suggests. The spines are immovable in some of them (known as burrfishes), but in members of the genus *Diodon* they are erected to stick straight out from the body as the fish swells. The largest of these fishes may grow to 70 cm (27.5 in) or more in length – enough to make most predators think twice when they encounter it fully inflated.

Triggerfishes and filefishes have a defensive ploy along similar lines. They are able to raise a spine – the first ray of the dorsal fin – which is locked upright by the second spine of the fin, acting as a trigger. At the same time they can lower a bone to distend their pelvic

BELOW **Porcupinefishes, such as the yellowspotted burrfish** *Cyclichthys spilostylus*, **generally forage at night.**

LEFT **Orangestriped triggerfish** *Balistapus undulatus*.

Toxic puffers

Many puffers produce the strong nerve poison tetrodotoxin, stored primarily in the liver and certain other organs. Despite this, they are held to be a delicacy, particularly in Japan where they are served up as *fugu*, but eating them without skilled preparation can prove fatal. Some puffers release a toxic glandular secretion from their skin, repelling predators. The guineafowl puffer *Arothron meleagris* does this, as do smaller members of the family, the sharpnose puffers (often known as tobies). A few species of filefishes have taken advantage of the puffers' defence, by mimicking them. In the best-known case, the harmless blacksaddle filefish *Paraluteres prionurus* gains protection by bearing a remarkable resemblance to the conspicuously banded toby *Canthigaster valentini*, with which it swims. Juveniles of the blacksaddle coralgrouper *Plectropomus laevis* are also thought to mimic that species of puffer.

An allied family, the trunkfishes or boxfishes (Ostraciidae), can also secrete a toxin – ostracitoxin – from the skin. It is specific to the family. They also have a bony casing and while their combination of defences deters most predators, some large pelagic fishes have been known to eat juveniles and even adults.

ABOVE **The black-saddled toby *Canthigaster valentini* is protected by its toxin.**

TOP **Whitespotted filefish** *Cantherhines macrocerus*.

BOTTOM **Smooth trunkfish** *Lactophrys triqueter*.

Nest-guarding triggerfish

Even the largest triggerfishes are generally shy, but there are occasions when their behaviour is quite the reverse. When guarding the nesting site, the titan triggerfish Balistoides viridescens is distinctly aggressive towards divers. Not only does the fish drive them out of its territory but it continues to give chase, and sometimes bites quite severely. As the fish may be up to 75 cm (30 in) in length and has a penchant for cracking open the toughest shellfish, it is perhaps not surprising that divers have learnt to be more wary of a brooding triggerfish than of the average reef shark! The picassofish Rhinecanthus aculeatus, a common triggerfish of shallow lagoons, is considerably smaller but it has been reported to attack reef walkers when it is nest-guarding. Filefishes do not have this aggressive streak, although they may tend their nests. Puffers also lay eggs on the sea bed, but are not known to guard them, while porcupinefishes and trunkfishes swim up to spawn in open water.

region. As with puffers, their apparent size is increased, and if retreat is still prudent, they have the means to wedge themselves securely in a refuge. In addition, the small inflator filefishes of the genus *Brachaluteres* can inflate their bodies in the manner of puffers.

Filefishes generally skulk around branching corals and can change their colour pattern to blend in, hoping to escape notice and avoid conflict (see also p. 84). Even the largest species, the scrawled filefish, behaves in this manner; it may grow to 1 m (3 ft) or more, but it is timid and, although it occurs throughout the tropics, it is not seen very often. The name 'filefish', alternatively 'leatherjacket', refers to the tough, rough-textured skin of these fishes, a feature of triggerfishes and puffers too.

Rigid-bodied trunkfishes are incapable of inflating, but they are well protected by a casing of bony plates, often giving them a strongly angular shape. They move slowly,

impeded by their armour and unable to bend. The plates are polygonal, as can be seen clearly in many species, for example the honeycomb cowfish *Acanthostracion polygonius* of the Western Atlantic. (Cowfishes are trunkfishes that have a pointed projection above each eye, resembling a pair of small horns.)

ABOVE **Detail of the bony plates encasing a trunkfish – the honeycomb cowfish *Acanthostracion polygonius*.**

Bottom-dwelling fishes

The best way to conserve energy is to lie still, and many bottom-dwelling fishes have perfected this to a fine art. Most of them are predatory, but catching food is not a problem. On a coral reef, the small fishes, crustaceans and other animals on which they feed are abundant and, sooner or later, will pass right in front of them. Lying out on the reef to take advantage of this stream of traffic carries the risk of being conspicuous, not only to potential prey but also to large and dangerous predators. Bottom-dwelling fishes have evolved various adaptations to overcome this problem – chiefly in their camouflaged appearance and their ability to remain motionless for long periods of time. Some

BELOW **Many bottom-dwelling fishes, such as the devil scorpionfish** *Scorpaenopsis diabolus*, **are well camouflaged.**

perch on corals or rocks, but many prefer the patches of coral rubble or sand that form a significant part of the reef environment. The neutral colours of most of these fishes, marked with speckles and blotches, attract scant attention in those surroundings. The majority of them are reluctant to swim, and move sluggishly only when disturbed, settling again nearby. Some flatfishes, however, are more active and skim speedily over the sea bed for some distance.

All these fishes appear to have much in common, so similar are they in temperament and colour pattern – but are they related? In fact they are far more diverse than you might imagine. The obvious traits they share are superficial devices to suit a sedentary lifestyle on the reef bed, and they do not indicate the lineages from which their owners have arisen.

The fishes wearing this corporate uniform are classified in a number of different families, and although some of them are closely related – scorpionfishes and flatheads for example – others, such as gobies and lizardfishes, are not even in the same superorder.

Scorpionfishes and their relatives

Scorpionfishes (family Scorpaenidae) are well disguised by their rugged, rock-like outlines and projecting flaps of skin that give a whiskery appearance. They rely on not being noticed, to the extent that they can be approached and even touched before they break cover. Such interference is inadvisable – tropical scorpionfishes are venomous, particularly lionfishes and stonefishes (see box, p. 69). Many members of the family live

TOP **One of the colour varieties of the leaf scorpionfish** *Taenianotus triacanthus*.

BOTTOM **A flathead or crocodilefish of the genus** *Cociella*.

on coral reefs, while others are found in temperate waters. One feature they have in common is a large spiny head with a characteristic bony ridge on the cheek.

Several species, such as the spotted scorpionfish *Scorpaena plumieri* in the Caribbean, flash a warning when alarmed, displaying startling colour patterns on the inner side of the pectoral fins as they move away. Then, when they settle and fold their fins, the predator is likely to lose track of the rubble-coloured creatures. Another species, the decoy scorpionfish *Iracundus signifer*, has fin markings for predatory reasons – the front end of its dorsal fin resembles a tiny fish, acting as a lure when the fin is waggled. Among the most bizarre are the rare scorpionfishes of the genus *Rhinopias*, which are covered with tassels and resemble a clump of seaweed or perhaps the feather-like arms of a crinoid. Scorpionfishes are generally sturdy but one of the smaller members of the family, the leaf scorpionfish *Taenianotus triacanthus*, is more delicate in build. It sways from side to side, with its narrow body and tall dorsal fin mimicking a drifting leaf or seaweed. Waspfishes (in the rarely seen family Tetrarogidae) behave in a similar fashion.

Flatheads

Flatheads (family Platycephalidae) have a longer, flattened version of the scorpionfish body-plan, but with two dorsal fins instead of one that is notched. The facial features of certain species – eyes bulging on top of a ridged head with long snout and large mouth – warrant the sometimes-used name 'crocodilefish'. The illusion is enhanced by the presence of alternating pale and dark markings round the edge of the face; these are tooth-like to the human eye but are merely part of the camouflage. These fishes can be seen lying on sandy areas or rubble patches on coral reefs in the Indo-Pacific, but not in the Atlantic.

Venom

The venom glands of scorpionfishes and lionfishes are situated at the base of the spines of the dorsal, anal and pelvic fins. When a spine punctures the skin of another animal, a groove along the spine allows the venom to flow into the wound. If the incautious animal survives the wound, it will learn to avoid another encounter.

The stonefish *Synanceia verrucosa* of the Indo-Pacific is the most dangerous fish on the reef, with venom matched in toxicity only by that of a few close relatives less associated with coral sites. Wounds from the stonefish are unbearably painful and occasionally fatal. Even wearing shoes may not give protection because the strong, stout spines can pierce a rubber sole. *Synanceia verrucosa* is larger than most other scorpionfishes; it is particularly bulky and may grow to about 35 cm (14 in) in length. Even so, the fish often escapes notice unless the ground is inspected closely; with its rugged warty skin it looks just like a chunk of coral rubble.

ABOVE **The stonefish *Synanceia verrucosa* is notorious for its venom.**

Lionfishes

Lionfishes are classified as the subfamily Pteroinae of the scorpionfishes and are sometimes called turkeyfishes. Unlike typical members of the scorpionfish family, lionfishes are boldly coloured, although their striped pattern appears less conspicuous amongst branching corals and other life on a crowded reef. Their fins set them apart from most other scorpionfishes. The dorsal fins are divided into separate long rays or spines, and the pectoral fins are enormous and sometimes also divided. They are found in the Indo-Pacific, typically hiding under ledges for most

RIGHT **Lionfish** *Pterois miles*.

of the day, but the two largest species (the lionfishes *Pterois volitans* and *Pterois miles*, which may reach 50 cm [20 in]) are likely to be seen out in the open, hovering a little way above the reef.

Although scorpionfishes and their relatives live on the sea bed, they release their spawn in open water either with the eggs embedded in a gelatinous mass or singly. Open-water spawning is the practice of many bottom-dwelling fishes, but others, such as gobies and blennies, lay their eggs on the sea bed.

Lizardfishes

The facial expression of a lizardfish (family Synodontidae) is distinctly reptilian – the wide mouth is usually slightly open, exposing one or more rows of numerous slender, sharply-pointed teeth. A lizardfish has a long slim body, which it props up on its pelvic fins,

tilting its pointed face slightly upwards for a good view of the surroundings. Sandperches of the genus *Parapercis* have a similar habit, but they are not in the same family as lizardfishes and they have a different dentition. Lizardfishes frequently use low rocks or corals as lookout posts, but some prefer to hide, burying all except the head in the sand. The sand diver *Synodus*

ABOVE **The variegated lizardfish *Synodus variegatus* perches on the reef.**

BELOW **Speckled sandperch *Parapercis hexophthalma*.**

71

intermedius, a common species in the Caribbean, often adopts the latter mode, although it will also sit on rock.

Flatfishes

Various species of flatfishes, which belong to different families, live on coral reefs. Most are sandy-coloured on the upper side, which blends with the ground they lie on, but some are more obvious. Peacock flounders (*Bothus mancus* and *Bothus lunatus*) are marked with blue rings, while the zebra sole and unicorn sole (*Zebrias zebra* and *Aesopia cornuta*, respectively) have a series of bold blackish stripes across their pale bodies. Some flatfishes bury themselves in the sand to escape detection, but certain peacock soles, as well as moses soles of the genus *Pardachirus* deter predators – even sharks – by secreting a bitter toxin.

Frogfishes

Frogfishes (family Antennariidae) are highly camouflaged, strongly resembling the sponge-covered rock or coral surfaces most of them favour as resting places. Their many colour varieties match a range of backgrounds, and even if they are seen they are likely to be mistaken for sponges or other distasteful invertebrates. Limb-like pectoral fins enable frogfishes to manoeuvre their lumpy bodies, but they do not move around much – nor do they need to. They entice their prey with a fleshy lure at the tip of the first spine of the dorsal fin, waggled above a mouth that can engulf a meal as large as themselves. Batfishes of the related family Ogcocephalidae, which are occasionally seen on coral reefs in the western Atlantic, also have limb-like fins and a lure on the front of the head.

RIGHT **Lefteye flounders of the genus *Bothus* are found on sandy areas of the reef.**

RIGHT **The shortstripe goby** *Elacatinus chancei* **dwells on sponges.**

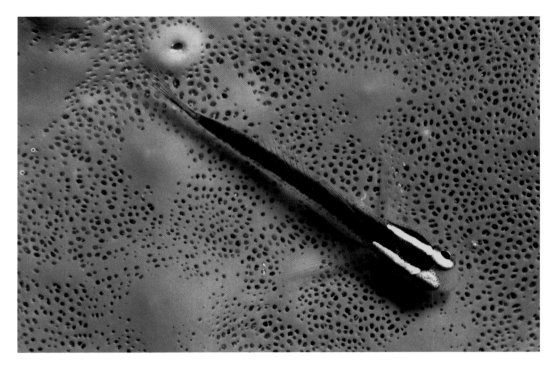

Gobies

Gobies are the largest family of marine fishes (the Gobiidae), with nearly 2000 species, almost half of which live on coral reefs. They are generally small, but tiniest of all is *Trimmatom nanus*, barely 1 cm (0.4 in) in length. Most gobies have no swimbladder, and do not need one because they typically rest on the sea bed, on a choice of surfaces varying from species to species. They usually have the pelvic fins joined together to form a suction disc, which helps to prevent those that perch on rock or coral from being swept away by currents (see p. 87). The pelvic disc is also useful on fine mud and similar smooth sediments, conditions where it has been speculated that gobies may have originated and where many live today. Others occupy more coarse-grained territory, generally with rocks and stones nearby for cover. Gobies often live in burrows or bury themselves in the sand for protection, and certain species of the genus *Trimma*, many of which are tiny and secretive, shelter in caves.

Most gobies eat small invertebrate animals on the sea bed, but some feed on plankton, and a few have a specialized diet. Several Caribbean gobies in the genus *Elacatinus* act as cleaners, removing and eating ectoparasites from other fishes, in a similar manner to the cleaner wrasse of the Indo-Pacific (p. 51). When they are not actively performing this service they rest on sponges or mounds of corals. Unlike most bottom-dwelling fishes, they need to be visible, and they have contrasting stripes along the body to alert customers for cleaning. (See also pp. 15, 28.)

Living in a burrow

A sandy flat is dangerously open territory, devoid of the countless ready-made hiding places that other parts of the reef offer, but it is easy to burrow into and many sand-dwellers do just that. The practice is common in gobies, and the burrow-dwelling species tend to be brighter in colour than their drab, camouflaged neighbours that have no such retreat. Burrows are used as nesting sites for spawning, and non-burrowing gobies seek out an alternative receptacle – an empty shell, as is the habit of certain blennies (family Blenniidae), or a sheltered corner of a ledge. The male goby then guards the eggs.

A pair of gobies often shares a burrow, sometimes taking turns at keeping watch and feeding. The twinspot or crab-eye goby *Signigobius biocellatus* (in the West Pacific, from the Philippines to Vanuatu) is one such species. One fish remains alert while the other sieves mouthfuls of sand through its gill rakers to extract small organisms. This goby has an unusual response to threatening situations. Its size of up to 10 cm (4 in) is only moderate for a goby but its markings are highly conspicuous. Two large eye-spots startle predators when the fish raises its dorsal fins, and the fish makes jerky movements, giving the impression of a sidling crab.

A number of goby species, usually living in pairs, share a burrow in the sand with prawns of the genus *Alpheus*, known as snapping or pistol shrimps because of the loud noises they can make with their pincers. The prawn (or shrimp) is blind or nearly so, but it makes itself useful by digging the burrow and shovelling out the shifting sand, while the goby keeps post as a sentry by the entrance.

LEFT **A twinspot or crab-eye goby *Signigobius biocellatus* displays its eye-spots to intimidate predators.**

Other small sand-burrowers include jawfishes (family Opistognathidae), some of which use fragments of coral rubble or shell to line their burrows. A jawfish either rests half out of its more-or-less vertical hole, or hovers above it, almost standing on its tail fin, alert and at the ready to make a quick tail-first retreat. Jawfishes have unusual breeding behaviour, with the males incubating the eggs in the mouth. In other circumstances 'housecleaning' is carried out by mouth and the debris is spat clear of the burrow.

A few morays (notably the ribbon eel) and eels of other families burrow in sand. Snake eels (family Ophichthidae) spend much of the day buried but when they are seen out and about their snake-like appearance is remarkable. This is particularly so for the banded snake eel *Myrichthys colubrinus*, which is easily mistaken for the venomous sea snake *Laticauda colubrina* that it mimics. Garden eels, reef-dwelling members of the conger family (Congridae), are extremely timid and keep their hind ends tucked well down in their burrows while they reach up to catch plankton drifting in the water above. Unlike other eels, they live in colonies, in which the evenly spaced ranks of long stalk-like bodies sticking up out of the ground resemble 'gardens'.

Blennies and related fishes

There are about 380 species of blennies (family Blenniidae), most of which live on coral reefs. They are slender-bodied, not unlike gobies in appearance but without scales. Blennies are typically small, around 12 cm (5 in) or less in length, and they have varied feeding styles. Some are omnivorous, but many reef-dwellers fall into the following two groups.

Combtooth blennies

In one camp are the primarily herbivorous combtooth blennies. These have a wide mouth with a single row of tiny sharp teeth, which enable them to scrape fine filamentous algae from rock surfaces. One member, the leopard blenny *Exallias brevis* in the Indo-Pacific, feeds on living coral tissues. Combtooth blennies are common in shallow parts of the reef and some, the rockskippers, are intertidal.

BELOW **The Maldives blenny *Ecsenius minutus* is a member of the herbivorous combtooth blennies.**

Sabretooth blennies

Another blenny group is predatory, as its name 'sabretooth' implies. These blennies have a smaller mouth but with stronger teeth, and they have large canine teeth in the lower jaw, although these are used for defence, not for feeding. Aggressive members of this group, known as fangblennies, nip bits of skin and scales off other fishes, and even bite divers occasionally. Their fish victims learn to shun them, but fangblennies have a few tricks that make it difficult. Their main ploy involves mimicking species that are harmless or even beneficial. The cleaner mimic or false cleaner *Aspidontus taeniatus* is well known for masquerading as the Indo-Pacific cleaner wrasse *Labroides dimidiatus* (but giving a treacherous nip instead of providing the expected cleaning services), and juveniles of the bluestriped fangblenny *Plagiotremus rhinorhynchos* have the same strategy. Some other species of *Plagiotremus* mimic poison-fang blennies of the genus *Meiacanthus*, which are not aggressive but are not preyed upon because they possess venomous grooved fangs to ward off attack. Species of *Meiacanthus* may be found on the reef bed but they are more often seen hovering off the bottom to feed on zooplankton. Other mimics mingle with them, not only blennies but also juveniles in unrelated families of fishes – certain spinecheeks of the genus *Scolopsis* and cardinalfishes of the genus *Cheilodipterus*.

Blenny-like fishes

Blenny-like fishes include the triplefins (family Tripterygiidae), which are unique among coral fishes in that they have three dorsal fins.

There are numerous species of these tiny, scaled fishes but the only ones commonly seen are those few that rest in the open, on large sponges or sometimes on coral mounds. Males and females are sometimes differently coloured, especially at breeding times. One species has a more unusual sexual difference: males of the rhinocerus triplefin *Helcogramma rhinoceros* have a projecting upper lip resembling a horn.

Also related to true blennies are the scaled blennies (family Labrisomidae), many of which live in the Western Atlantic. They are generally camouflaged in a variety of reef habitats, but the diamond blenny *Malacoctenus boehlkei* escapes attack by hiding amongst the tentacles of pink-tipped anemones *Condylactis gigantea*. Like the anemonefishes of the Indo-Pacific region, this Caribbean fish appears to be immune to the stinging cells of its protector. Tubeblennies (otherwise known as flag blennies) have sometimes been classed in the same family as the scaled blennies, but are more widely considered to be a distinct family (the Chaenopsidae). The front end of the dorsal fin is prominent in several tubeblennies, and the males display characteristic markings when they raise it.

Inside the reef

Coral reefs are riddled with holes. Some are natural crevices; others are made and later vacated by worms, molluscs and other invertebrate animals that bore into the coral rock. Small holes, such as empty worm tubes, are ideal homes for tiny blennies. The blenny's head sticks out of the hole and swivels round to catch sight of any nearby activity, bobbing back in when prudent but rarely for long.

OPPOSITE **A bluestriped fangblenny *Plagiotremus rhinorhynchos* peeps out from a hole in the reef. Juveniles are coloured differently and are aggressive mimics of cleaner wrasses.**

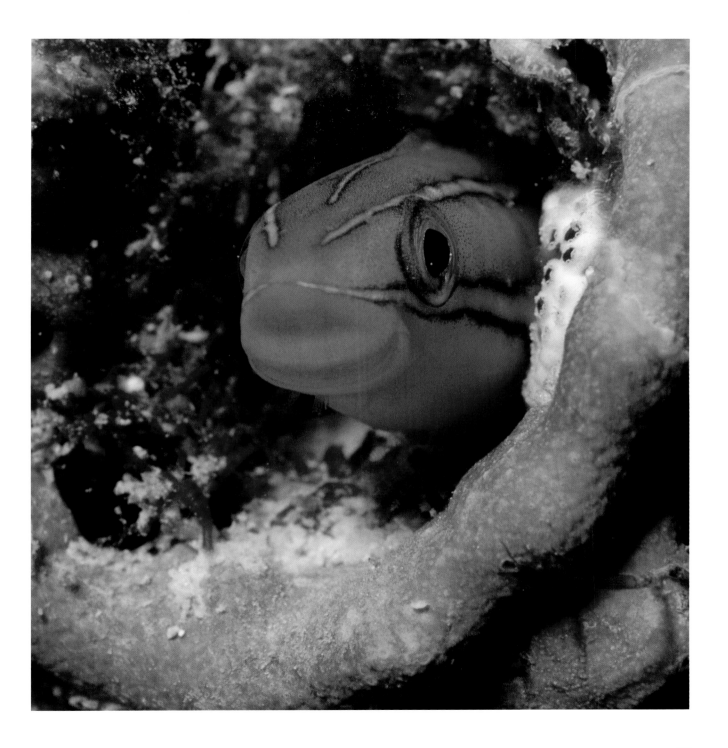

Some of those most often seen are the midas blenny *Ecsenius midas*, and the bluestriped fangblenny *Plagiotremus rhinorhynchos* (which has the alternative common name of tube-worm blenny). Both are Indo-Pacific species.

In the Caribbean it is mainly tubeblennies that occupy worm tubes or similar holes made by clams, for example. These small predatory fishes dart out to grab shrimps or other minute crustaceans but only two species venture far from their holes, the arrow blenny *Lucayablennius zingaro* and the wrasse blenny *Hemiemblemaria simulus*. The arrow blenny is a very slender fish with a long pointed mouth and snout. Its feeding method is to drift with its tail curled, but ready to flick into action, propelling the fish forward to snatch a goby or similar small prey. The wrasse blenny has very different behaviour. It swims into midwater to feed, hovering with bluehead wrasses *Thalassoma bifasciatum* and rapidly adapting its bright yellow colour to an almost indistinguishable match of the yellow phase of the wrasse. The wrasse is left unharmed by predatory fishes because it is a cleaner, and the mimic enjoys the same protection.

Eels in the reef

Tiny fishes are not the only ones to use holes in the reef – morays (family Muraenidae) do too, although they require something much larger than a worm hole. Morays are long-bodied eels that like to move about inside the reef, winding through passages to hunt their prey. When they are not hunting, morays look out from the main entrance of their homes,

ABOVE **The spotted moray *Gymnothorax moringa* is a common Caribbean species.**

with smaller holes nearby perhaps revealing a glimpse of the lower body or tail sliding by as the moray shifts its position. Some are timid and withdraw into the darker recesses when approached. Morays rarely attack divers without provocation, although they will bite readily in defence, but one species – the masked moray *Gymnothorax breedeni* – which occurs around various islands in the Indo-Pacific is notably aggressive. Fortunately it is unlikely to inflict serious damage. Most fishes are wary of the morays' sharp teeth, so a moray mimic escapes predation. The comet *Calloplesiops altivelis*, a member of the prettyfin family (Plesiopidae), has a spotted pattern similar to that of the whitemouth moray *Gymnothorax meleagris*. When the comet is alarmed it heads into a crevice, leaving its tail sticking out, with a false eye-spot creating the illusion of a moray's head.

Occasionally a moray can be seen out in the open, streaking across the reef – an impressive sight if it is a large individual. Several species reach a length of 1 m (3 ft) or more. On coral reefs there are two outstanding examples: in the Caribbean, the green moray *Gymnothorax funebris*, which is capable of growing to well over 2 m (6.5 ft), and in the Indo-Pacific, the giant moray *Gymnothorax javanicus*, reaching 3 m (10 ft) and a reputed weight of 70 kg (154 lb). Well-grown individuals can dispatch large prey, which can make a gruesome spectacle. A tussle observed between a green moray and a trumpetfish *Aulostomus maculatus* resulted in the latter being sliced in two like a stick of celery, before its remains, still making rhythmic but ineffectual movements, were dragged into the lair and gradually swallowed.

The commoner morays are rather drab in the main, but a few rarer or more secretive species are spectacular in appearance. The dragon moray *Enchelycore pardalis* has a ferocious air, with curved jaws exposing long pointed teeth. Its rear nostrils are extended as tubes, and its colour pattern is striking – orange with black-ringed white blotches. The sand- or rubble-dwelling ribbon eel *Rhinomuraena quaesita* is also remarkable. The black juveniles develop into blue-and-yellow males and may then change sex to become plainer yellow females. Adult ribbon eels have pronounced nose flutes, the nostrils forming large flared flaps, and it has been speculated that these may act as lures for prey capture.

Hiding amongst corals

Some of the fishes that live on the reef bed, as well as various others that lie low, have adopted hiding amongst corals as a mode of life. Many resemble their coral habitats but even those that are not camouflaged are unobtrusive, so stealthy is their behaviour. A number of them, such as hawkfishes and trumpetfishes, are ambush hunters, while others hide primarily to protect themselves from predators. Their coral haunts may shield them also from water currents. Out in the open many of these fishes, especially pipefishes and their relatives, are weak swimmers and vulnerable.

Hawk-like watchers

Hawkfishes (family Cirrhitidae) usually perch amongst the outer branches of corals, holding themselves in place by means of the thickened lower rays of their pectoral fins. From their chosen vantage points, they remain still and unobserved while on the lookout, ready to dart out and grab their prey of small fishes or crustaceans passing by. Most species are found on true corals, or on various soft corals and black corals but a few dwell on sponges.

The longnose hawkfish *Oxycirrhites typus* is typically found on sea fans or on black corals. Its body is criss-crossed with red lines,

LEFT **A fire coral serves as a look-out post for a blackside hawkfish** *Paracirrhites forsteri*.

creating a patchwork appearance that, from a distance, merges with the branching structure of the coral. Like certain other hawkfishes it may vary in colour according to habitat, and individuals on straw-coloured sea fans are generally paler than those on red ones.

The hawkfish's sharp eyes miss little and it reacts warily when approached, employing a game of hide-and-seek as a defensive measure. The tangle of branches of a black coral bush make a perfect hideout for the fish to retreat into but it is more exposed on a sea fan, against the flat meshwork of tiny branches. There are always gaps in the coral's structure though, and the fish can dart through one of these to the other side. A large predator is then unable to follow the hawkfish directly and must swim around the sea fan, often a span of a couple of metres, by which time the small fish has found another suitable perch on the sea fan with a bolt-hole close by.

ABOVE **The longnose hawkfish *Oxycirrhites typus* is camouflaged against a sea fan.**

ABOVE **Tassel-like filaments on the dorsal fin of the dwarf hawkfish *Cirrhitichthys falco* are characteristic of the hawkfish family.**

ABOVE **Pixy hawkfish *Cirrhitichthys oxycephalus*.**

The hawkfish family

Most of the 32–35 species of hawkfishes are found on Indo-Pacific reefs, and only one (the redspotted hawkfish *Amblycirrhitus pinos*) lives in the Caribbean Sea. Male hawkfishes are territorial and usually live with a small harem of females. However, a few species depend on rich coral growth to maintain a harem and are monogamous in poorer situations. Hawkfishes spawn in open water, at dusk, and the larval phase is pelagic but adults generally dwell on the reef bed. One species, the swallowtail hawkfish *Cyprinocirrhites polyactis*, behaves differently, swimming above the reef to feed on plankton. Hawkfishes are mostly about 15 cm (6 in) or less in length, and have characteristic filaments, often tassels, at the tips of their dorsal fin spines.

RIGHT **The trumpetfish** *Aulostomus maculatus* **often mimics long-branched gorgonian corals.**

BELOW **The mouth and snout of a trumpetfish** *Aulostomus chinensis* **form a tube, which can expand greatly to suck in prey.**

Coral mimics

Trumpetfishes of the genus *Aulostomus* use their resemblance to corals to devastating effect, as far as their prey is concerned. The fishes skulk by gorgonian corals, the long stick-like body hovering almost vertically alongside the branches. The pipette-like snout sucks in small fishes and shrimps quickly and efficiently, before these victims have any chance to react to impending danger. Trumpetfishes can change colour rapidly, and the single Caribbean species *Aulostomus maculatus* typically assumes a chequered camouflage pattern all over its body. Like their allies the cornetfishes *Fistularia*, and certain other fishes, trumpetfishes sometimes ride alongside or above a larger fish, using its bulk to hide against.

Razorfishes or shrimpfishes (see box on p. 86) have perfected the technique of hovering

vertically in the shelter of corals, although juveniles, and sometimes also adults, prefer to stay close to long-spined sea urchins for protection. Their unusual behaviour is fascinating to watch. Small groups of the razorfish *Aeoliscus strigatus* may be seen in colonies of sea whips, one of their favoured habitats. The fish hang with head pointing down, and when you swim towards them they back off slowly, still holding the same pose and moving in perfect synchrony with one another. When the group is on the outskirts of the sea whips, the fish switch instantaneously to horizontal swimming mode, to make a quick break for the safety of the next patch of sea whips.

Several species of filefishes live among the branches of gorgonians and other soft corals, where they drift slowly, typically with head pointing slightly down. Filefishes can change colour; the camouflage of smaller species is sometimes enhanced by flaps and tassels of skin. You need to be sharp-eyed to see the minute leatherjacket *Rudarius minutus* and the radial leatherjacket *Achreichthys radiatus* in the Indo-Pacific, or the slender filefish *Monacanthus tuckeri* in the Western Atlantic, all of which are secretive fishes, usually only a few centimetres in length.

One of the most spectacular coral mimics is the harlequin ghost pipefish *Solenostomus paradoxus*, which is typically found swimming head down beside gorgonians, or occasionally other soft corals or black coral bushes. It also lives amongst the fronds of seaweeds, the feather-like arms of crinoids and the spines of sea urchins. Although the fish is widespread in the Indo-Pacific it is

encountered rarely, and then only in tranquil environments where there is little water movement. Whisker-like filaments and tassels cover it from the tip of the snout to the fan-shaped tail, disguising its body perfectly. Despite its ornate markings, which are often red and white, its outline blends almost imperceptibly with the profusion of life on a coral outcrop.

ABOVE **Razorfish (shrimpfish)** *Aeoliscus strigatus* **in typical pose amongst sea whips.**

Pipefishes and seahorses

Pipefishes and seahorses often live amongst seaweeds or sea grass, but some are found on coral reefs. Although they are generally inconspicuous and well-camouflaged, if you examine a ledge or crevice closely you might see the narrow ribbon-like body of a pipefish when it moves. Some reef-dwelling pipefishes are on living corals, and a few are associated with particular coral species. Several of these are known in the Indo-Pacific, including the coral pipefish *Siokunichthys nigrolineatus*, which is found hiding among the anemone-like tentacles of certain mushroom corals. A closely related species is known on soft corals, and the pugheaded or eel pipefish *Bulbonaricus brauni* occurs on organ-pipe coral and some true corals.

Seahorses *Hippocampus* are another rare find on corals. The few that are associated with them live on sea fans (gorgonian corals), curling the tail around the branches to maintain position – like certain pipefishes they have no tail fin and are poor swimmers. Resting quietly against the sea fan, they might be taken for part of the coral growth. Pygmy seahorses *Hippocampus bargibanti* and related species are amazingly good mimics of the sea fans they live on; the large knobs protruding from their bodies look remarkably like the polyps of the particular corals they choose. This, and their small size (rarely more than 2 cm [1 in]) makes them especially difficult to find. However, they may be numerous locally: up to 28 pairs have been recorded on a single sea fan in Papua New Guinea.

LEFT **A minute pygmy seahorse *Hippocampus* sp. (closely related to *H. bargibanti*), is barely visible against the branches of the sea fan *Subergorgia mollis*.**

ABOVE **Harlequin ghost pipefish** *Solenostomus paradoxus*.

Pipefishes and their relatives

Some of the fishes that hide amongst corals are members of the same group, the Gasterosteiformes. This order of strange-looking and often very elongated fishes includes several families, of which the pipefish and seahorse family (Syngnathidae) is by far the largest, with over 200 species, although not all of them live on coral reefs. Other families in the order have fewer species – the razorfishes (Centriscidae) have about four species, and the ghost pipefishes (Solenostomidae) and the trumpetfishes (Aulostomidae) have only three currently named species apiece, although there are further species of ghost pipefishes as yet undescribed. Also in the Syngnathiformes are two small families that have no coral-mimicking members, the cornetfishes (Fistulariidae) and the seamoths (Pegasidae), and the sticklebacks and certain other fishes not of coral reefs.

A feature of the Syngnathiformes is a long tubular snout, often fluted at the tip, for sucking in tiny animals from the water. Pipefishes, seahorses and razorfishes have another feature in common: they are encased in bony plates that make their bodies more rigid. Seahorses and pipefishes have long been renowned for their strange mode of breeding, in which the male receives the eggs from the female and incubates them on his body or tail, sometimes in a specially developed pouch; in seahorses this pouch is on the belly. In ghost pipefishes the female broods the eggs in her enlarged pelvic fins.

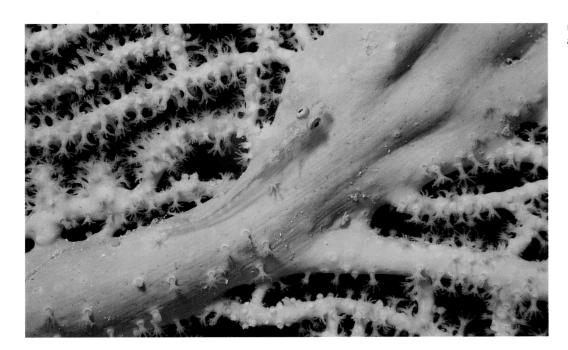

LEFT **A tiny goby lies against a sea fan.**

Coral gobies

The large and varied family of gobies includes a number of species that dwell on living corals. Some are conspicuous, notably the Caribbean cleaner gobies that perch on top of prominent corals (see p. 73). These fishes have contrasting colours that draw attention to themselves and attract the clients they service. Others hide: Indo-Pacific gobies of the genera *Gobiodon* and *Paragobiodon* are not usually camouflaged, but they rest so far down amongst the branches of stony corals that they are rarely seen.

Dwarfgobies such as *Bryaninops* species are often very well camouflaged. They inhabit sea whips, sea fans and other soft corals where, in a similar manner to the longnose hawkfish, they dart round to the other side of the coral branch when a predator approaches.

Even before they hide, the tiny fishes are not easy to detect because they align their slender bodies with the branches. They are usually a good colour match for the coral, and have bars and other disruptive markings that make the outline of the body less obvious, or else they are transparent.

Gobies have a ventral sucker, a particular advantage for those on vertical branches in exposed situations. It means that they can safely cling there while swept by water currents that bring minute plankton for both fish and coral to feed on. These gobies stay with one coral, or others in the immediate vicinity, for life. Usually one male associates with two or more females. As with hawkfishes and certain other fishes, if the male dies one of the females will change sex to replace him.

Nocturnal fishes

A marked change takes place on the reef as daylight fades. Fishes that have been feeding during the day hurry to their roosts and another company – the night feeders – emerges. Many nocturnal fishes belong to three families: bigeyes (Priacanthidae), squirrelfishes and soldierfishes (Holocentridae), and cardinalfishes (Apogonidae). Although they are not closely related they have certain features in common, which each family has developed independently to fit its members for night activity. Many of them are red, and they have large eyes to aid night vision, most

obviously in bigeyes. The mouth is large but the teeth are generally small, suited to preying mainly on the planktonic animals that are plentiful after dark, when they rise up from deeper water.

By day they live in sheltered sites, often hiding in caves or other shaded places, but when darkness falls they move out into open water to feed. A few live at great depths, well below the limits of reef-building corals – the deep-water soldierfish *Ostichthys kaianus* is known from about 300 to over 600 m (from 984 to 1968 ft), and the longfinned bullseye

LEFT **Longspine squirrelfish *Holocentrus rufus*.**

RIGHT **Snappers of the genus** *Pinjalo*.

Cookeolus japonicus (a member of the bigeye family) has been found down to about 400 m (1312 ft) in various parts of the tropics. The bullseye is the largest fish in the three families, reaching 60 cm (24 in) or more in length, while among more typical reef fishes the sabre squirrelfish *Sargocentron spiniferum* may grow to about 45 cm (18 in). At the other end of the size range are the cardinalfishes, delicate in appearance and some no more than 2–3 cm (around 1 in) in length.

Bigeyes

In addition to the general features of nocturnal fishes, a bigeye has a large head and a strongly downturned mouth with a projecting lower jaw that, together with its outsize eyes, give it an air of sour watchfulness. Bigeyes are often solitary but sometimes form shoals, particularly at deeper sites. Although typically deep red, their colour varies and some are blotched. They can change colour rapidly, and may become silver if they move into the open during the day, but they do not do this in unison. As the shoal swims over a coral head some fish may turn pink or silver while others next to them are still red, so that they seem to be a disparate group.

There are 18 species in the bigeye family, the majority of which occur in the Indo-Pacific. Four species live in the Caribbean and western Atlantic but only two of them live at depths where they are likely to be observed. These are the glasseye (or glasseye snapper) *Heteropriacanthus cruentatus*, on shallow reefs throughout the tropics, and the bigeye

LEFT **Squirrelfish**
Holocentrus ascensionis.

Priacanthus arenatus, on deeper reefs (but within scuba-diving range). Bigeyes have a varied diet of larger plankton that includes shrimp-like crustaceans along with the larvae of fishes, crabs and cephalopods, and they may supplement this with a little daytime feeding also.

Squirrelfishes and soldierfishes

The squirrelfishes and soldierfishes are a larger family than the bigeyes and include about 70 species, far more of which occur in the Indo-Pacific than in the Caribbean and Western Atlantic. They are all spiny, but squirrelfishes differ from soldierfishes in that they have a large, somewhat venomous, spine projecting backwards from the preopercle,

whereas this is very small or absent in soldierfishes. One feature that is immediately noticeable about them all is their large coarse scales, which are much more conspicuous than those of bigeyes.

They sometimes have the unusual habit of swimming upside down as they follow the walls around a spacious cave or reach the ceiling of an overhang. When they forage at night, crustaceans form the major part of their diet, but while squirrelfishes generally catch crabs and shrimps on the reef, soldierfishes hover above in the water column after crab larvae and other plankton. Most members of this family are able to make clicking sounds and other noises in communication with each other, when they are courting or when they are alarmed.

Cardinalfishes

The largest group of nocturnal fishes is the cardinalfish family, in which there are 200 or more species, most of which are found on or near coral reefs. Because they are small, commonly less than 10 cm (4 in), and sometimes secretive, new species are still being discovered. Many cardinalfishes have some red colouring, as their name implies, but others do not. The Caribbean species are generally reddish, whereas the Indo-Pacific representatives display a wider range of colours. Their colours may instead be brown, yellow, black, and white, sometimes

LEFT **Many nocturnal fishes, such as the bigeye** *Priacanthus hamrur*, **are red.**

Seeing red?

Red is a dominant colour in night-active coral fishes, and also in many nocturnal shrimps and crabs, but it is not seen in the dark – red appears black then instead. Moreover, the eyes of many nocturnal fishes may be poorly adapted for colour vision. Thus, richly coloured animals are cloaked in darkness, an advantage when they are out in the open at night, whether they are stalking prey or avoiding becoming a predator's next meal. Their day-time refuges are also dark enough to conceal the colour. Caves and other recesses are obvious places in which to hide, but deep water is equally effective. Water absorbs light, with a strong bias towards the red end of the spectrum. Even at relatively shallow sites, around 6 m (19.5 ft), reds start to diminish, although other colours will still appear bright and the general scene may be bathed in sunlight. Below 30 m (98 ft) the scene is gloomier and red is no longer visible.

semi-transparent and often with stripes on their long oval-shaped bodies, from the head to the base of the tail. Some cardinals are solitary, others gather by day in small groups or in large schools, sometimes around branching corals or bushy soft corals. They often hover almost motionless as if suspended in the water, some feeding a little, but most waiting patiently until dusk signals the start of their nightly forage on small animals, such as copepods and other crustaceans from the plankton.

The large mouth typical of nocturnal fishes is put to a most unusual use by cardinals – they incubate their eggs there. The responsibility usually falls to the male, his throat swelling with the egg-mass he has received from the female. When he opens his mouth he reveals an amazing sight because the mass may contain well over 100 eggs. Mouth-brooding is rare in coral reef fishes; it is better known in cichlids, a large group of freshwater fishes. After hatching, larval cardinalfishes usually drift as plankton in the same manner as other coral fish larvae; the duration of this stage is up to several weeks for cardinals.

Strange relationships

Some cardinalfishes have extraordinary day-time hideouts, which offer these vulnerable fishes greater protection. Various small species in the Indo-Pacific can be found amongst the long spines of sea urchins, such as *Diadema* species. The sea urchin cardinalfish *Siphamia versicolor* is one of these, while two closely related species live among the spines of the crown-of-thorns starfish. They are easily

LEFT **Orange-lined cardinalfish *Archamia fucata* often gather in large groups.**

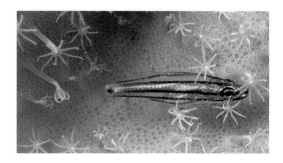

LEFT **A five-lined cardinalfish *Cheilodipterus quinquelineatus* shelters against a soft coral, but sometimes these fish take refuge among the spines of sea urchins.**

overlooked, being only a few centimetres in length, and are likely to be more common than reports have suggested.

In the Caribbean, one cardinalfish has an even more bizarre partnership. Known as the conchfish, *Astrapogon stellatus* commonly lives with the queen conch and hides in the mollusc's mantle cavity. Certain other Caribbean cardinals, in the genus *Phaeoptyx*, live in tubular sponges or with starfish. The sawcheek cardinalfish *Apogon quadrisquamatus* lives with sea anemones, but unlike Indo-Pacific anemonefishes, it is not immune to the stinging tentacles and risks damage to itself as the price it must pay for protection against predators.

Light organs

A few fishes are able to emit light, which is produced by symbiotic luminous bacteria in specific parts of their bodies. Flashlightfishes (family Anomalopidae) have a light organ beneath each eye. These have various uses: as lures and to locate small plankton for food; for communication; and to confuse predators. The fishes cannot turn the light off but they can hide it from view by various means, to perform a 'disappearing act'. Flashlightfishes of the genus *Photoblepharon* can cover the organ with a skin flap, as if blinking, while *Anomalops* species can rotate the organ, and *Kryptophaneron* species combine the two methods.

Flashlightfishes shun the light and live in very deep or dark places by day, moving up to shallower levels, where divers may see them, only on moonless nights. Steep reef faces (dropoffs) with caves are favoured locations. At night, divers may see small bright lights that move about in midwater, suddenly switch off, then reappear in unexpected places. The scattered lights seem eerily disembodied, but if you shine the beam of a torch or flashlight on one you can sometimes catch a fleeting glimpse of a small plain fish retreating hastily from the light.

Pineapplefishes (family Monocentridae) operate similar luminescent organs, either one or two, but they have them in a different place – on the lower jaw. The common name of these rarely-seen fishes refers to their strange appearance, yellow, bony-plated and spined. A few species of cardinalfishes, some sweepers, and the slipmouths (family Leiognathidae), use bacteria in their gut to produce luminescence.

LEFT **The light organ of the flashlightfish *Photoblepharon* sp. is conspicuous beneath the eye.**

Other night-time feeders

Sweepers (family Pempheridae) have a similar lifestyle to that of most other nocturnal fishes, sheltering during the day and feeding above the reef on zooplankton at night. They too have large eyes, but not the large mouth, nor much evidence of the red colour of many of their associates. Instead, they are transparent or dull bronze until a light is shone on them, when these small, oval fishes gleam in silver or copper hues. Sweepers are much the same size as cardinals, but the body is more compressed. They are sometimes seen in mixed groups with cardinals during the day but more often they form schools of single species, at the entrance of a cave or sometimes inside, or else in the shade of a coral head. There are a reasonable number of sweepers living on reefs in the Indo-Pacific and in the Western Atlantic. They may be quite conspicuous on shallow sites, particularly those of the genus *Parapriacanthus*, emboldened by the solidarity of belonging to dense schools.

Various other fishes feed at night. There are larger hunters such as snappers, and grunts or sweetlips (see p.100–102). Caribbean drums (family Sciaenidae) venture out from under ledges and find crabs, shrimps and polychaete worms to eat, and soapfishes (in the grouper and seabass family) do much the same in the Indo-Pacific.

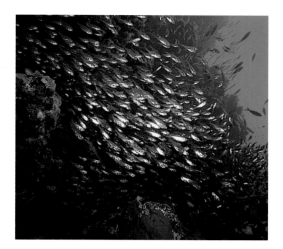

LEFT **Sweepers** *Parapriacanthus guentheri* **gather in dense swarms by day but disperse to feed at night.**

LEFT **Sweepers of the genus** *Pempheris* **are deep-bodied.**

LEFT **The spotted drum** *Equetus punctatus* **is secretive by day but hunts invertebrate animals at night.**

Hunters

Many of the larger fishes on or around coral reefs are hunters, either of other fishes or of active invertebrates such as squid, cuttlefish, octopuses and crabs. Some hunt at night but much of the predation on fishes takes place at dusk and dawn, when smaller fishes are moving to or from their refuges and are particularly vulnerable. Strategies vary: morays and groupers remain near their places of shelter on the reef bed and capture prey that passes, while sharks and jacks go after their victims in high-speed chases. The term 'coral fish' hardly seems appropriate for such aggressive predators, and they are more often visitors than true reef-dwellers, but a number of them are drawn to the rich source of food that a coral reef offers.

Sharks

At the top of the food chain are the sharks – the big predators. They tend to hunt more by night, and certain species rest on sand or coral rubble for much of the day, either in caves or out in the open. When seen in such situations, reef sharks (most of which are members of the requiem family,

OPPOSITE **The scalloped hammerhead shark *Sphyrna lewini* is an open-water species seen occasionally around reefs. It is not usually aggressive, but certain other hammerheads can be dangerous.**

BELOW **The gray reef shark *Carcharhinus amblyrhynchos*, as in other requiem sharks, is streamlined for rapid pursuit of prey.**

Carcharhinidae) are generally wary and stir from the bottom if they are approached, but nurse sharks (family Ginglymostomatidae) are sluggish and reluctant to move off.

Sharks are admirably efficient hunters: strong swimmers, with muscular jaws that give formidable biting power, and a dentition of renown. Typical sharks – the fish-eaters – have sharp pointed teeth, narrow ones for seizing and piercing their prey or the familiar triangular blades for cutting. Close examination of the latter reveals a serrated edge, which helps to sever outsize prey into chunks that can be swallowed. Other species, such as nurse sharks, have teeth that are more flattened, for crushing shellfish, a feature they share with rays.

Dangerous sharks – reality or myth?

While almost any shark may bite under provocation, sharks generally prefer to avoid confrontation. However, a few species can attack unpredictably. The most dangerous shark on coral reefs is the rarely seen tiger shark *Galeocerdo cuvier* – it has a particular liking for turtles and sea birds but it will consider eating almost any animal, and has been responsible for numerous human fatalities. 'Shark' is an emotive word, provoking mixed reactions of fear and fascination, but anyone who has seen a shark cruising underwater in its natural environment is inspired with awe and admiration for the magnificent animal. Most sharks, even large ones, are surprisingly wary and difficult to approach closely. The dangers that sharks present are overstated, but sharks themselves may be endangered by human activities, particularly by the trade in shark fins.

Rays

Like sharks, rays are cartilaginous fishes, but with bodies shaped like flattened discs or diamonds as opposed to the narrow elongated form of sharks. Rays are grouped in several families, including stingrays (Dasyatidae, and the round stingrays Urolophidae), eagle rays (Myliobatidae) and manta rays (Mobulidae). Stingrays flush their prey of shellfish out of the sand, but most of their day is spent lying on the sand, exposed or covered by it. Either way they would seem to be well protected by one or more venomous spines near the base of the tail, but they are no match for hammerhead sharks, which come in from deeper water to feed on stingrays and other fishes. A hammerhead shark can detect a ray that is buried in the sand, and uses its head to hit the ray and pin it to the ground before it can escape. Among the most common reef rays are the southern stingray *Dasyatis americana*, in the Caribbean, and the bluespotted ribbontail ray *Taeniura lymma*, in the Indo-Pacific. The latter ray, patterned as its name implies, is one of the few conspicuously coloured species. Most rays and sharks are sombre in hue, mottled for camouflage on the sea bed, or a plainer steely grey that merges with the distant gloom in open water.

Barracudas

Barracudas (family Sphyraenidae) rival sharks as voracious predators. The long silver body is poised, like a javelin with jaws, to dart at and snatch a browsing parrotfish or even an unwary jack. The great barracuda *Sphyraena*

LEFT **The great barracuda *Sphyraena barracuda* is a powerful-jawed predator.**

BOTTOM LEFT **A covering of sand hides much of the southern stingray *Dasyatis americana*.**

barracuda, which may reach 2 m (6.5 ft) in length, is solitary, but other species form schools.

Harmless giants

Some of the greatest spectacles to be witnessed on coral reefs are provided by two enormous visitors from the open ocean, the whale shark *Rhincodon typus* and the manta ray *Manta birostris*. The whale shark is the largest fish in the world, sometimes exceeding 12 m (29 ft) in length. The manta ray is smaller but impressive none the less, with a fin span that can be well over 6 m (19.5 ft). Despite their size, it is safe to swim alongside them, giving snorkellers and divers an occasional

thrilling encounter at sites throughout the tropics. Both fishes are plankton-feeders, swimming near the surface with their wide mouths open to take in their catch and filter it from the water. The manta ray has the advantage of a pair of flaps at the front of its head to channel plankton towards its mouth. The whale shark can use its huge gape to create a suction force, drawing water in as it pulls its head beneath the surface. Fishes are ingested along with the plankton; these are usually small ones but whale sharks have been known to swallow mackerel and tuna whole. Sharks are slow-growing, long-lived animals, and whale sharks are no exception, capable of living for around 60 years.

ABOVE **The whale shark** *Rhincodon typus* **swims almost at the surface with mouth agape to engulf plankton.**

Snappers and grunts

Snappers (Lutjanidae) and grunts (Haemulidae) are two generally similar families of fishes, although probably not as closely related as was once thought. They are sizable fishes, many species growing to 30 cm (12 in) or more and occasional species reaching 1 m (3 ft), and they are conspicuous and sometimes abundant on coral reefs. By day some of them, especially Caribbean grunts, gather in groups or schools, but at night the majority disperse to feed. Their prey are mainly crustaceans and other invertebrate animals, although the juvenile fishes and some adults feed on plankton by day.

Grunts

There are 150 species of grunts, which occur throughout the tropics but which are particularly common on shallow reefs in the Caribbean and western Atlantic. Many of them are brightly coloured, featuring yellow and silver with stripes or spots. In the Indo-Pacific (where thick-lipped grunts are often called sweetlips or, occasionally, rubberlips) their juveniles are strikingly different and boldly patterned. One of the most notable is

Sound-producers

Grunts are known to make grunting noises by grinding their pharyngeal teeth (situated in the throat), and a wide range of fishes are capable of sound production in communication with their own species or others. These sounds may serve a variety of purposes. They have been reported in courtship rituals, for example, but loud calls are chiefly defensive, either as territorial, warning-off signals, or as alarm calls when confronted by a major threat. The swim bladder usually serves as an amplifier and, together with its musculature, is specialized for resonation in drums and croakers (family Sciaenidae), used to make clicks by squirrelfishes, and used also by some angelfishes and groupers. Certain triggerfishes, pufferfishes, and their relatives, make noises by rubbing either their teeth or their fin spines, and often also involve the swim bladder to amplify the noise.

Fishes that 'chatter' to each other may pay a price. It has been discovered that bottlenose dolphins home in on grunts and other sound-producers, and eat them. Similarly, predators such as sharks, jacks and snappers listen out for the noises that fishes make, often involuntarily, in the course of feeding and swimming.

the young harlequin sweetlips *Plectorhinchus chaetodonoides*, which has large white blotches ringed with black as opposed to the speckles of the adult. Its behaviour when very young is even more remarkable; it swims with an undulating motion in the manner of a flatworm or a sea slug that has been dislodged from the reef, possibly mimicking these unpalatable animals. Differences between juveniles and adults are less marked in Caribbean grunts, but most species have some markings – a black stripe and a spot at the base of the tail – which they lose on reaching maturity.

BELOW **The oriental sweetlips *Plectorhinchus vittatus* is an Indo-Pacific member of the grunt family.**

RIGHT **Giant trevally**
Caranx ignobilis **chasing**
fusiliers.

RIGHT **Giant trevally**
Caranx ignobilis **chasing fusiliers.**

of smaller fishes react to by closing ranks. The pace quickens when jacks hunt; they single out a victim and then attack at lightning speed using their protrusible jaws to capture their prey.

When large predators move into hunting mode the change of mood on the reef is palpable. Shoaling fishes that were milling about the corals a moment before shoot out from the reef edge in unison, like a shock wave from an explosion. Jacks such as the bluefin trevally *Caranx melampygus* sometimes work as a pack to break up the schools of their prey and scatter them, making it easier to home in on individuals. Other species commonly associated with coral reefs include shoaling horse-eye trevally *Caranx latus* and bigeye trevally *Caranx sexfasciatus*. Some jacks are large (several species may exceed 1 m (3 ft)) and, like snappers, grunts, and various groupers, they are heavily fished.

Glossary

cartilaginous fishes fishes with a skeleton of cartilage

coral head a prominent clump of coral

coral polyp the individual animal of a coral colony

coralline algae algae with a hard calcified structure

crinoid a feather star, member of the echinoderms (starfish, sea urchins, and their relatives)

dimorphic having two genetically determined morphological types or forms

dorsal toward the back (in a fish this is the upper part of the body)

drop-off a nearly vertical slope of the outer or seaward portion of a reef, a 'wall'

El Niño a complex set of changes in the water temperature in the Eastern Pacific, producing a warm current

endemic native to, and restricted to, a particular geographic region

genus a category of classification comprising one or more related species

gill rakers bony projections on the gill arch (the bony structure which supports the gill tissues of a fish)

gizzard a muscular grinding chamber of the digestive tract of certain animals

gorgonian horny coral

hydroid member of a group of animals (the order Hydroida) related to sea anemones, corals, and jellyfish

lunate used to describe the tail fin of a fish in which the margin is curved as in a new moon, or is sickle-shaped

mantle (of a mollusc) part of the body wall that secretes the shell, covers the visceral mass, and encloses the mantle cavity

Müllerian mimicry a type of mimicry in which a number of inedible species have similar appearance, typically warning coloration, which a predator learns to avoid

nematocysts stinging cells of cnidarians or coelenterates (the group of animals to which sea anemones, corals, jellyfish and hydroids belong)

order a rank between class and family in the hierarchy of the classification of organisms

pectoral fin the fin usually present behind the gill opening on either side of a fish's body

pedicellariae minute pincer-like appendages on the body surface of echinoderms (starfish, sea urchins, and their relatives)

pharynx region of the digestive tract behind the mouth cavity

photosynthesis the biochemical process by which green plants synthesize carbohydrates from carbon dioxide and water, using sunlight as the energy source

plankton animals and plants that drift in the ocean's currents

pharyngeal teeth patches of teeth on the gill arches of bony fishes, sometimes modified to form a grinding apparatus or pharyngeal mill

preopercle the 'cheek' region of a fish, the anterior bone of the gill cover

reef crest the highest point on a reef, before it drops down to the reef edge and the seaward slope into deep water

reef wall a nearly vertical slope of the outer or seaward portion of a reef

scleractinia present day true or stony corals

scute an external bony plate or enlarged scale

sea fan gorgonian (horny coral) with a fan-like growth form

sea whip gorgonian (horny coral) with a whip-like growth form

symbiosis living together – a close association or partnership between animals or plants, usually of unrelated species

table coral a table-like growth form of certain corals of the genus *Acropora*

tribe a rank between family and genus in the hierarchy of the classification of organisms

truncate with a blunt or square end

zooplankton animals that drift in the oceans' currents

zooxanthellae single-celled algae that live symbiotically within the cells of reef-forming corals and some other marine animals

Index

Bold refers to main references. *Italics* refer to captions. ***Bold italics*** refer to feature boxes.

Further information

Further reading

Caribbean Reef Fishes, John E. Randall. T.F.H., Neptune City, NJ, 3rd edn.,1996.

Coral Reef Fishes, Indo-Pacific & Caribbean, Ewald Lieske & Robert Myers. Collins Pocket Guide, HarperCollins, London, 1994 (reprinted 1996).

Encyclopedia of Fishes, John R. Paxton & William N. Eschmeyer (Consultant Eds). Academic Press Inc., 2nd edn., 1998.

Fishes of the Great Barrier Reef and Coral Sea, John E. Randall, Gerald R. Allen & Roger C. Steene. University of Hawaii Press, and Crawford House Press, 2nd edn., 1997.

Indo-Pacific Coral Reef Field Guide, Gerald R. Allen & Roger Steene. Tropical Reef Research, Singapore, 1994.

Micronesian Reef Fishes, Robert F. Myers. Coral Graphics, Guam, 3rd edn., 1999.

Reef Fish Identification; Florida Caribbean Bahamas, Paul Humann. New World Publications, Inc., Jacksonville, 2nd edn., 1994.

The Diversity of Fishes, Gene Helfman, Bruce B. Collette & Douglas E. Facey. Blackwell Science, 1997.

The Ecology of Fishes on Coral Reefs, Peter F. Sale (Ed). Academic Press Inc., 1991.

The Greenpeace Book of Coral Reefs, Sue Wells & Nick Hanna. Blandford, London, 1992.

Internet resource

FishBase: A Global Information System on Fishes
http://www.cgiar.org/iclarm/fishbase/

Picture credits

Photographs by Linda Pitkin, with the exception of the following:
Oxford Scientific Films (Peter Parks) 17
Brian Pitkin 28 (bottom), 35, 50 (bottom), 63 (bottom), 74, 89, 93 (top), 95 (middle)
Planet Earth Pictures (Peter Rowlands) 94

Acknowledgements

The author would like to thank colleagues in the Department of Zoology at The Natural History Museum, London, for invaluable advice and critical comments on the manuscript; Peter Millar and Helmut Debelius, for helpful advice and comments; Brian Pitkin for his help and support during the preparation of this book; and finally, I.W. Urquhart and Connex staff for help in retrieving proofs left on a train.